Ben Stacy Jerrik (Ed.)

E-readiness

Ben Stacy Jerrik (Ed.)

E-readiness

Economist Intelligence Unit, Information and communications technology, United Nations Public Administration Network

Part Press

Contents

Articles

References

E-readiness

E-Readiness is the ability to use information and communication technologies (ICT) to develop one's economy and to foster one's welfare.

There are several benchmarking indices at the macro (also called global, universal, etc.) level, e.g., those calculated by the UNPAN, World Bank, Economist Intelligence Unit etc.

Because what appear on the macro level can hide wide heterogeneity among organizations (educational institutions, government departments, etc.) local areas (cities, towns, etc.) individuals (female, individuals with disabilities, etc.) in digital access, a micro level more detailed benchmarking is suggested to compute sub-measures for networking, applications, web-accessibility and readiness (NAWAR).

E-Readiness indices at the macro level are constructed primarily for ranking countries, facilitating comparisons between countries and over time. They can also be used to track the global digital divide, i.e. the gap between countries that have access to ICT and those that do not (mainly because of differences in income, education, etc.).

NAWAR is constructed primarily to measure how ICT is actually put to work for development. For example, NAWAR is concerned with the gap between humans with respect to natural / assistive access to ICT in e-business environments, i.e., whether organizations have assistive systems (e.g. Braille keyboards and printers, one-handed keyboards, annotated websites for screen reading software, etc.) and whether organizational cultures adopt green computing. More importantly, because NAWAR is concerned with how ICT is actually put to work for development, attention is given to change in the level of activity, i.e. the move from e-readiness to impact in e-business environments.

Economist Intelligence Unit e-readiness rankings

Each year, in cooperation with the IBM Institute for Business Value, the Economist Intelligence Unit produces a ranking of e-readiness across countries, based on six pillars of e-readiness: connectivity & technology infrastructure, business environment, social & cultural environment, legal environment, government policy & vision and consumer & business adoption.

[1] [2]

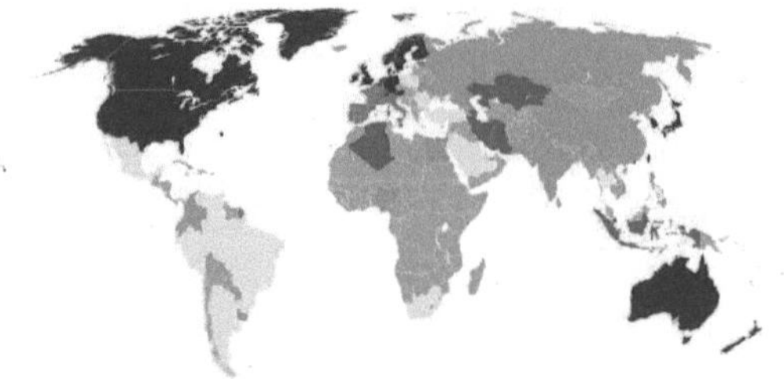

World map showing the e-readiness scores 8.000 - 8.999 7.000 - 7.999 6.000 - 6.999 5.000 - 5.999 4.000 - 4.999 3.000 - 3.999 No Data

In the 2009 e-readiness rankings, global e-readiness fell partly due to fall in the global economy in the later part of 2008. In addition, the report started to cover ICT usage, which better reflected how countries are using ICT effectively. [3]

Rank 2009	2008	2007	2006	Country	e-readiness score (out of 10) 2009	2008	2007	2006
1	5	1	1	Denmark	8.87	8.83	8.88	9.00
2	3	2	4	Sweden	8.67	8.85	8.85	8.74
3	7	8	6	Netherlands	8.64	8.74	8.50	8.60
4	11	12	11	Norway	8.62	8.60	8.35	8.35
5	1	2	2	United States	8.60	8.95	8.85	8.88
6	4	9	8	Australia	8.45	8.83	8.46	8.50
7	6	6	13	Singapore	8.35	8.74	8.60	8.24
8	2	4	10	Hong Kong	8.33	8.91	8.72	8.36
9	12	13	9	Canada	8.33	8.49	8.30	8.37
10	13	10	7	Finland	8.30	8.42	8.43	8.55
11	16	14	14	New Zealand	8.21	8.28	8.19	8.19
12	9	5	3	Switzerland	8.15	8.67	8.61	8.81
13	8	7	5	United Kingdom	8.14	8.68	8.59	8.64
14	10	11	14	Austria	8.02	8.63	8.39	8.19
15	22	22	19	France	7.89	7.92	7.77	7.86
16	19	17	23	Taiwan	7.86	8.05	8.05	7.51
17	14	19	12	Germany	7.85	8.39	8.00	8.34
18	21	21	16	Ireland	7.84	8.03	7.86	8.09
19	15	16	18	South Korea	7.81	8.34	8.08	7.90
20	20	20	17	Belgium	7.71	8.04	7.90	7.99
21	17	15	20	Bermuda	7.71	8.22	8.15	7.81
22	18	18	21	Japan	7.69	8.08	8.01	7.77
23	23	24	-	Malta	7.46	7.78	7.56	-
24	28	28	27	Estonia	7.28	7.10	6.84	6.71
25	26	26	24	Spain	7.24	7.46	7.29	7.34
26	25	25	25	Italy	7.09	7.55	7.45	7.14
27	24	23	22	Israel	7.09	7.61	7.58	7.59
28	27	27	26	Portugal	6.86	7.38	7.14	7.07
29	29	29	28	Slovenia	6.63	6.93	6.66	6.43
30	32	30	31	Chile	6.49	6.57	6.47	6.19
31	31	31	32	Czech Republic	6.46	6.68	6.32	6.14
32	38	41	38	Lithuania	6.34	6.03	5.78	5.45
33	30	32	29	Greece	6.33	6.72	6.31	6.42
34	35	33	30	United Arab Emirates	6.12	6.09	6.22	6.32
35	33	34	32	Hungary	6.04	6.30	6.16	6.14
36	36	39	36	Slovakia	6.02	6.06	5.84	5.65

				Country				
37	37	37	39	Latvia	5.97	6.03	5.88	5.30
38	34	36	37	Malaysia	5.87	6.16	5.97	5.60
39	41	40	34	Poland	5.80	5.83	5.80	5.76
40	40	38	39	Mexico	5.73	5.88	5.86	5.30
41	39	35	35	South Africa	5.68	5.95	6.10	5.74
42	42	43	41	Brazil	5.42	5.65	5.45	5.29
43	43	42	45	Turkey	5.34	5.64	5.61	4.77
44	49	46	43	Jamaica	5.33	5.17	5.05	4.67
45	44	44	42	Argentina	5.25	5.56	5.40	5.27
46	50	-	-	Trinidad and Tobago	5.14	5.07	-	-
47	48	48	44	Bulgaria	5.11	5.19	5.01	4.86
48	45	45	49	Romania	5.07	5.46	5.32	4.44
49	47	49	47	Thailand	5.00	5.22	4.91	4.63
50	53	52	54	Jordan	4.92	5.03	4.77	4.22
51	46	46	46	Saudi Arabia	4.88	5.23	5.05	5.03
52	58	53	51	Colombia	4.84	4.71	4.69	4.25
53	51	51	49	Peru	4.75	5.07	4.83	4.44
54	55	55	56	Philippines	4.58	4.90	4.66	4.41
55	52	50	48	Venezuela	4.40	5.06	4.89	4.47
56	56	56	57	China	4.33	4.85	4.43	4.02
57	57	58	55	Egypt	4.33	4.81	4.26	4.30
58	54	54	53	India	4.17	4.96	4.66	4.04
59	59	57	52	Russia	3.98	4.42	4.27	4.14
60	63	59	58	Ecuador	3.97	4.17	4.12	3.88
61	62	62	60	Nigeria	3.89	4.25	3.92	3.69
62	61	60	61	Ukraine	3.85	4.31	4.02	3.62
63	60	61	59	Sri Lanka	3.85	4.35	3.93	3.75
64	65	65	66	Vietnam	3.80	4.03	3.73	3.12
65	68	67	62	Indonesia	3.51	3.59	3.39	3.39
66	64	63	67	Pakistan	3.50	4.10	3.79	3.03
67	67	66	63	Algeria	3.46	3.61	3.63	3.32
68	70	69	65	Iran	3.43	3.18	3.08	3.15
69	66	64	64	Kazakhstan	3.31	3.89	3.78	3.22
70	69	68	68	Azerbaijan	2.97	3.29	3.26	2.92

See also

- Information and Communication Technologies for Development

References

[1]   "2007 EIU e-readiness rankings" (http://graphics.eiu.com/files/ad_pdfs/2007Ereadiness_Ranking_WP.pdf). Economist Intelligence Unit. 2007. .

[2]   "2008 EIU e-readiness rankings" (http://graphics.eiu.com/upload/ibm_ereadiness_2008.pdf). Economist Intelligence Unit. 2008. .

[3]   "2009 EIU e-readiness rankings" (http://graphics.eiu.com/pdf/E-readiness rankings.pdf). Economist Intelligence Unit. 2009. .

External links

- The Electronic Journal of Information Systems in Developing Countries: E-Readiness for Developing Countries (http://www.ejisdc.org/ojs2/index.php/ejisdc/article/viewFile/219/184)
- Center for International Development at Harvard University: Readiness for the Networked World — A Guide for Developing Countries (http://cyber.law.harvard.edu/readinessguide/)
- SchoolnetAfrica: E-Readiness as a Tool for ICT Development (http://www.schoolnetafrica.org/fileadmin/resources/E-readiness_as_a_tool.pdf)
- 2001 EIU e-readiness rankings (http://www.aph.gov.au/house/committee/jfadt/CentralEurope/ce_report/appendixi.pdf)
- 2002 EIU e-readiness rankings (http://unpan1.un.org/intradoc/groups/public/documents/APCITY/UNPAN010005.pdf)
- 2003 EIU e-readiness rankings (http://graphics.eiu.com/files/ad_pdfs/eReady_2003.pdf)
- 2004 EIU e-readiness rankings (http://graphics.eiu.com/files/ad_pdfs/ERR2004.pdf)
- 2005 EIU e-readiness rankings (http://graphics.eiu.com/files/ad_pdfs/2005Ereadiness_Ranking_WP.pdf)
- 2006 EIU e-readiness rankings (http://graphics.eiu.com/files/ad_pdfs/2006Ereadiness_Ranking_WP.pdf)
- Extensive bigliographic collection (http://ictlogy.net/bibciter/reports/types_categories.php?idcat=21) on e-Readiness

Economist_Intelligence_Unit

The **Economist Intelligence Unit** (**EIU**) is an independent business within The Economist Group.[1]

Official logo of the Economist Intelligence Unit.

Through research and analysis, EIU offers forecasting and advisory services to its clients. It provides country, industry and management analysis worldwide and incorporates the former Business International Corporation, a U.K. company acquired by the parent organization in 1986. It is particularly well known for its monthly country reports, five-year country economic forecasts, country risk service reports, and industry reports. The company also specialises in tailored research for companies that require analysis for particular markets or business sectors. 2006 marked the 60th anniversary of the Economist Intelligence Unit's inception.

The Economist Intelligence Unit also produces regular reports on the "liveability", and cost of living, of the world's major cities, which receive wide coverage in international news sources. The Economist Intelligence Unit's Quality-of-Life Index is another noted report.

Its current Editorial Director & Chief Economist is Robin Bew.

Economist Corporate Network

Economist Corporate Network is The DUI's business intelligence, advisory, briefing and networking service for senior executives. Its members benefit directly from insight on economic and business trends in emerging markets and from access to a wide network of peers and analysts.

CHAMPS

In November 2010 the Economist Intelligence Unit released the Access China White Paper profiling the economies of the top 20 emerging cities in China. The report coined the acronym CHAMPS (Chongqing, Hefei, Anshan, Maanshan, Pingdingshan and Shenyang).[2]

Democracy Index

In 2006 (with updates in 2008, 2010 and 2011) the Economist Intelligence Unit released The Democracy Index, an index compiled by examining the state of democracy in 167 countries, attempting to quantify this with an Economist Intelligence Unit Index of Democracy which focused on five general categories: electoral process and pluralism, civil liberties, functioning of government, political participation and political culture.[3]

Government Broadband Index (gBBi)

In January 2011, the Unit released the Government Broadband Index (gBBi) that assesses countries on the basis of government planning, as opposed to current broadband capability. With ambitious targets for both the speed and coverage of next-generation broadband networks, the developed countries of South-east Asia scored highest. According to the index Greece is the worst-performing country measured, owing to its relatively low coverage target and drawn-out deployment schedule. Greece also suffers due to the considerable size of its public-funding commitment as a percentage of overall government budget revenues, and because its plan does little to foment competition in the high-speed broadband market.[4]

See also

- E-readiness rankings
- Quality-of-Life Index
- Democracy Index
- CHAMPS

References

[1] Economist Intelligence Unit's web page "who we are" (http://www.eiu.com/public/who-we-are.aspx)

[2] THE RISE OF THE 'CHAMPS' - NEW REPORT MAPS BUSINESS OPPORTUNITY IN CHINA'S FASTEST GROWING CITIES (http://www.sourcewire.com/releases/rel_display.php?relid=60590)

[3] Democracy under Stress (http://www.sourcewire.com/releases/rel_display.php?relid=69243&hilite=)

[4] South Korea tops the EIU's inaugural government broadband index and Japan comes a close second (http://www.sourcewire.com/releases/rel_display.php?relid=62568)

External links

- Official website (http://www.eiu.com)
- Business Research website (http://www.businessresearch.eiu.com)
- CHAMPS White Paper (http://www.eiu.com/champs)
- Democracy Index: Democracy under stress (http://www.eiu.com/democracyindex2011)
- Full speed ahead: The government broadband index Q1 2011 (http://www.eiu.com/public/topical_report.aspx?campaignid=broadband2011)

Information_and_communications_technology

Information and communications technology or **information and communication technology,**[1] usually abbreviated as **ICT**, is often used as an extended synonym for information technology (IT), but is usually a more general term that stresses the role of unified communications[2] and the integration of telecommunications (telephone lines and wireless signals), computers, middleware as well as

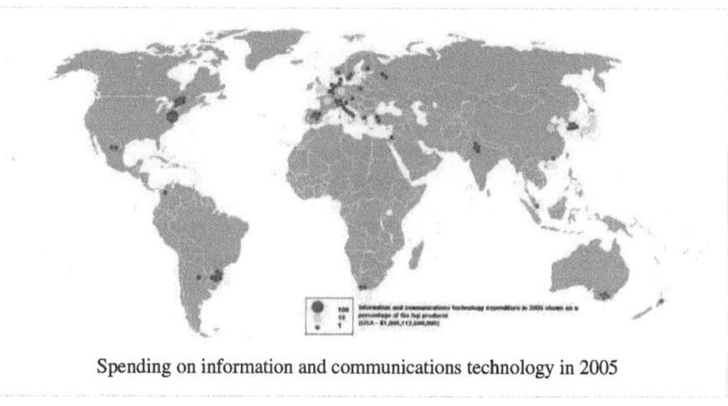

Spending on information and communications technology in 2005

necessary software, storage- and audio-visual systems, which enable users to create, access, store, transmit, and manipulate information. In other words, ICT consists of IT as well as telecommunication, broadcast media, all types of audio and video processing and transmission and network based control and monitoring functions.[3]

The expression was first used in 1997[4] in a report by Dennis Stevenson to the UK government[5] and promoted by the new National Curriculum documents for the UK in 2000.

The term *ICT* is now also used to refer to the merging (convergence) of audio-visual and telephone networks with computer networks through a single cabling or link system. There are large economic incentives (huge cost savings due to elimination of the telephone network) to merge the audio-visual, building management and telephone network with the computer network system using a single unified system of cabling, signal distribution and management. This in turn has spurred the growth of organizations with the term ICT in their names to indicate their specialization in the process of merging the different network systems.

Some of the dangers of ICTs include cyber-bullying, phishing as well as masquerading.

Trends and new concepts in the ICT roadway

ICT is often used in the context of "ICT roadmap" to indicate the path that an organization will take with their ICT needs[6] . It is also used as an overarching term in many schools, universities and colleges stretching from Information Systems/Technology at the organisational end through to Software Engineering and Computer Systems Engineering at the other[7] .

Standards

Standards are very important for ICT, since they define the language that enables the technology to understand each other. This is especially relevant because the key idea behind ICT is that information storage devices can communicate in media-frictionless manner with communication networks and computing systems. Open standards play a special role, as well as standards organizations such as the Telecommunications Industry Association in the United States and ETSI in Europe.

Campaigns and Projects

Recently, many companies are coming up with campaigns to promote ICT and how it can be utilised to protect nature. Wipro was a leading company that organized Earthian, wherein students came up with ideas to use technology to benefit society.

UNESCO's Information for All Program (IFAP) provides a platform for discussion of and action on ethical, legal and societal consequences of ICT developments.

See Also

- Global e-Schools and Communities Initiative
- ICT Development Index
- Information Age
- Information and Communication Technologies for Development
- Information and Communication Technologies for Environmental Sustainability
- Market Information Systems
- Mobile Web

References

[1] Sometimes used with *technologies* in the plural. Originally, only *information and communications technology* (with *communications* in the plural) was considered correct since ICT refers to *communications* (in the sense of a method, technology, or system of sending and receiving information, specifically telephone lines, computers, and networks), not *communication* (the act of sending or receiving information by speaking, writing, phoning, emailing, etc. or a message containing such information), and the older form (*information and communications technology*) is still the only one recorded in professionally edited reference works (e.g. Oxford Dictionaries Online (http://www. oxforddictionaries.com/view/entry/m_en_gb0398080#m_en_gb0398080), Computer Desktop Encyclopedia (http://encyclopedia2. thefreedictionary.com/ICT), Webopedia (http://www.webopedia.com/TERM/I/ICT.html), and Encarta® World English Dictionary (http://encarta.msn.com/encnet/features/dictionary/DictionaryResults.aspx?refid=561547268)) and preferred by many style guides (e.g. Editorial Style Guide of the Republic of South Africa (http://www.gcis.gov.za/resource_centre/guidelines/styleguide/ editorial_styleguide_2010.pdf). Nevertheless, the form *information and communication technology* is becoming increasingly common and is now used in about half the books that can be searched using Google Books and is for example also used by the International Telecommunication Union (http://www.itu.int/ITU-D/ict/publications/idi/2009/material/IDI2009_w5.pdf).

[2] Cloud network architecture and ICT - Unified communication(UC) technologies, integration of telecommunications, computers, middleware and the data systems that support, store and transmit UC communications between systems. (http://itknowledgeexchange.techtarget.com/

modern-network-architecture/cloud-network-architecture-and-ict/)

[3] http://foldoc.org/Information+and+Communication+Technology

[4] http://specials.ft.com/lifeonthenet/FT3NXTH03DC.html

[5] The Independent ICT in Schools Commission (1997) Information and Communications Technology in UK Schools, an independent inquiry. (http://web.archive.org/web/20070104225121/http://rubble.ultralab.anglia.ac.uk/stevenson/ICT.pdf) London, UK. Author: chair Dennis Stevenson

[6] http://www.microsoft.com/education/MSITAcademy/curriculum/roadmap/default.mspx

[7] "Information Technology and Software Engineering" (http://www.flinders.edu.au/science_engineering/csem/disciplines/itse/). . Retrieved 8 December 2011.

Further reading

- Caperna A., Integrating ICT into Sustainable Local Policies. ISBN13:9781615209293 (http://www.igi-global.com/bookstore/Chapter.aspx?TitleId=43194)
- Carnoy, Martin. " ICT in Education: Possibilities and Challenges (http://www.uoc.edu/inaugural04/eng/carnoy1004.pdf)." Universitat Oberta de Catalunya, 2005.
- " Good Practice in Information and Communication Technology for Education (http://www.adb.org/documents/guidelines/good-practice-in-ict-for-education/Good-Practice-in-ICT-for-Education.pdf)." Asian Development Bank, 2009.
- Grossman, G. and E. Helpman (2005), "Outsourcing in a global economy", Review of Economic Studies 72: 135-159.
- Oliver, Ron. " The Role of ICT in Higher Education for the 21st Century: ICT as a Change Agent for Education (http://elrond.scam.ecu.edu.au/oliver/2002/he21.pdf)." University, Perth, Western Australia, 2002.
- Walter Ong, Orality and Literacy: The Technologizing of the Word (London, UK: Routledge, 1988), in particular Chapter 4
- *Measuring the Information Society: The ICT Development Index* (http://www.itu.int/ITU-D/ict/publications/idi/2009/material/IDI2009_w5.pdf). International Telecommunication Union. 2009. pp. 108. ISBN 92-61-12831-9.

External links

- Information and Communication Technologies for Development and Poverty Reduction: The Potential of Telecommunications (http://www.ifpri.org/pubs/jhu/icttelecom.asp)
- ICT in Education - UNESCO (http://www.unescobkk.org/education/ict)
- Web 2.0 (http://www.oreillynet.com/pub/a/oreilly/tim/news/2005/09/30/what-is-web-20.html)
- Information Technology and the Dream of Democratic Renewal (http://willmedia.will.uiuc.edu/ramgen/CAS/MillerComm2008-01-23.rv)
- The Roadmap for Open ICT Ecosystems: A Guide for Policymakers and Technologists (http://cyber.law.harvard.edu/epolicy/)
- Cloud network architecture and ICT - Unified communication(UC) technologies, integration of telecommunications, computers, middleware and the data systems that support, store and transmit UC communications between systems. (http://itknowledgeexchange.techtarget.com/modern-network-architecture/cloud-network-architecture-and-ict/)

United_Nations_Public_Administration_Network

The **United Nations Public Administration Network** mission statement is to promote the sharing of knowledge, experiences and best practices throughout the world in sound public policies, effective public administration and efficient civil services, through capacity-building and cooperation among the United Nations Member States, with emphasis on south-south cooperation and UNPAN's commitment to integrity and excellence. It is organised by the UN Division for Public Administration, part of the UN Department of Economic and Social Affairs.

Uniqueness

With communication technologies advancing at a tremendous rate, electronic information centres, online research organizations and specialized e-networks are becoming increasingly common and duplicative. UNPAN is different. Its substantive capacity and specialized service combine to create a uniqueness that sets UNPAN apart from conventional web-based information sources. The one feature that is most unusual about UNPAN is its universality as an online public sector policy and management information and knowledge network. UNPAN offers:

- Access to regional experience in the practice of public policy development and management at the regional, national and local levels;
- Capacity-building and south-south cooperation in information and knowledge management;
- Ease of access to worldwide information in all areas of public sector policy and management;
- Demand-driven and interactive two-way provider of information and knowledge network.

UNPAN is a dynamic process and not a static outcome which aims at responding to the needs of its users and addresses their most critical areas of work.

Core Themes

Governance systems and institutions

Public service and management innovation

Social and economic governance

Public financial management

Knowledge systems and e-government

Services/Products

Online information services

Online training services

Online advisory services

Online conference services

Online directory services

See also

* United Nations

External links

* Official site [1]
    * UNPAN international partners [2]
* UN Division for Public Administration [3]

References

[1] http://www.unpan.org
[2] http://www.unpan.org/discover.asp
[3] http://www.unpan.org/dpepa.asp

World_Bank

World Bank	
	World Bank logo
Type	International organization
Legal status	Treaty
Purpose/focus	Crediting
Location	Washington, D.C., U.S.
Membership	188 countries [1] (IBRD) 170 countries (IDA)
President	Robert Zoellick (present president) Jim Yong Kim (elected on April 16, 2012, and assumes office on July 1, 2012)
Main organ	Board of Directors [2]
Parent organization	World Bank Group
Website	worldbank.org [3]

The **World Bank** is an international financial institution that provides loans[4] to developing countries for capital programs.

The World Bank's official goal is the reduction of poverty. According to the World Bank's Articles of Agreement (As amended effective 16 February 1989) all of its decisions must be guided by a commitment to promote foreign investment, international trade and facilitate capital investment.[5]

The World Bank differs from the World Bank Group, in that the World Bank comprises only two institutions: the International Bank for Reconstruction and Development (IBRD) and the International Development Association (IDA), whereas the former incorporates these two in addition to three more:[6] International Finance Corporation (IFC), Multilateral Investment Guarantee Agency (MIGA), and International Centre for Settlement of Investment Disputes (ICSID).

History

The World Bank is one of five institutions created at the Bretton Woods Conference in 1944. The International Monetary Fund, a related institution, is the second. Delegates from many countries attended the Bretton Woods Conference. The most powerful countries in attendance were the United States and United Kingdom, which dominated negotiations.[8]

Although both are based in Washington, D.C., the World Bank is traditionally headed by a citizen of the United States while the IMF is led by a European citizen.

Lord Keynes (right) and Harry Dexter White, the "founding fathers" of both the World Bank and the IMF. Seen here at the Bretton Woods conference, where plans were laid to launch the two institutions.[7]

1944–1968

From its conception until 1965 the bank undertook a relatively low level of lending. Fiscal conservatism and careful screening of loan applications was common. Bank staff attempted to balance the priorities of providing loans for reconstruction and development with the need to instill confidence in the bank.[9]

Bank president John McCloy selected France to be first recipient of World Bank aid; two other applications from Poland and Chile were rejected. The loan was for US$250 million, half the amount requested and came with strict conditions. Staff from the World Bank monitored the use of the funds, ensuring that the French government would present a balanced budget and give priority of debt repayment to the World Bank over other governments. The United States State Department told the French government that communist elements within the Cabinet needed to be removed. The French Government complied with this diktat and removed the Communist coalition government. Within hours the loan to France was approved.[10]

The Marshall Plan of 1947 caused lending by the bank to change as many European countries received aid that competed with World Bank loans. Emphasis was shifted to non-European countries and until 1968, loans were earmarked for projects that would enable a borrower country to repay loans (such projects as ports, highway systems, and power plants).

1968–1980

From 1968 to 1980, the bank concentrated on meeting the basic needs of people in the developing world. The size and number of loans to borrowers was greatly increased as loan targets expanded from infrastructure into social services and other sectors.

These changes can be attributed to Robert McNamara who was appointed to the presidency in 1968 by Lyndon B. Johnson.[11] McNamara imported a technocratic managerial style to the Bank that he had used as United States Secretary of Defense and President of the Ford Motor Company.[12] McNamara shifted bank policy toward measures such as building schools and hospitals, improving literacy and agricultural reform. McNamara created a new system of gathering information from potential borrower nations that enabled the bank to process loan applications much faster. To finance more loans, McNamara told bank treasurer Eugene Rotberg to seek out new sources of capital outside of the northern banks that had been the primary sources of bank funding. Rotberg used the global bond market to increase the capital available to the bank.[13] One consequence of the period of poverty alleviation lending was the rapid rise of third world debt. From 1976 to 1980 developing world debt rose at an average annual rate of 20%.[14] [15]

In 1980, the World Bank Administrative Tribunal was established to decide on disputes between the World Bank Group and its staff where allegation of non-observance of contracts of employment or terms of appointment had not been honored.[16]

1980–1989

In 1980, A.W. Clausen replaced McNamara after being nominated by US President Jimmy Carter. Clausen replaced a large number of bank staffers from the McNamara era and instituted a new ideological focus in the bank. The replacement of Chief Economist Hollis B. Chenery by Anne Krueger in 1982 marked a notable policy shift at the bank. Krueger was known for her criticism of development funding, as well as of third world governments as rent-seeking states.

Lending to service third world debt marked the period of 1980–1989. Structural adjustment policies aimed at streamlining the economies of developing nations were also a large part of World Bank policy during this period. UNICEF reported in the late 1980's that the structural adjustment programs of the World Bank were responsible for the "reduced health, nutritional and educational levels for tens of millions of children in Asia, Latin America, and Africa".[17]

1989–present

From 1989, World Bank policy changed in response to criticism from many groups. Environmental groups and NGOs were incorporated in the lending of the bank in order to mitigate the effects of the past that prompted such harsh criticism.[18]

Traditionally, and due to tacit agreement between the United States and Europe, the U.S. has always chosen the President of the World Bank. In 2012, for the first time, there are two candidates nominated for the presidency of the World Bank who are not from the United States.

On 23 March 2012, U.S. President Barack Obama announced that the United States would nominate Jim Yong Kim as the next President of the Bank.[19]

Criteria

Many achievements have brought the Millennium Development Goals (MDGs) targets for 2015 within reach in some cases. For the goals to be realized, six criteria must be met: stronger and more inclusive growth in Africa and fragile states, more effort in health and education, integration of the development and environment agendas, more and better aid, movement on trade negotiations, and stronger and more focused support from multilateral institutions like the World Bank.

The World Bank headquarters in Washington, DC

1. **Eradicate Extreme Poverty and Hunger**: From 1990 through 2004, the proportion of people living in extreme poverty fell from almost a third to less than a fifth. Although results vary widely within regions and countries, the trend indicates that the world as a whole can meet the goal of halving the percentage of people living in poverty. Africa's poverty, however, is expected to rise, and most of the 36 countries where 90% of the world's undernourished children live are in Africa. Less than a quarter of countries are on track for achieving the goal of halving under-nutrition.

2. **Achieve Universal Primary Education**: The number of children in school in developing countries increased from 80% in 1991 to 88% in 2005. Still, about 72 million children of primary school age, 57% of them girls, were not being educated as of 2005.

3. **Promote Gender Equality**: The tide is turning slowly for women in the labor market, yet far more women than men- worldwide more than 60% – are contributing but unpaid family workers. The World Bank Group Gender Action Plan was created to advance women's economic empowerment and promote shared growth.

4. **Reduce Child Mortality**: There is some what improvement in survival rates globally; accelerated improvements are needed most urgently in South Asia and Sub-Saharan Africa. An estimated 10 million-plus children under five died in 2005; most of their deaths were from preventable causes.

5. **Improve Maternal Health**: Almost all of the half million women who die during pregnancy or childbirth every year live in Sub-Saharan Africa and Asia. There are numerous causes of maternal death that require a variety of health care interventions to be made widely accessible.

6. **Combat HIV/AIDS, Malaria, and Other Diseases**: Annual numbers of new HIV infections and AIDS deaths have fallen, but the number of people living with HIV continues to grow. In the eight worst-hit southern African countries, prevalence is above 15 percent. Treatment has increased globally, but still meets only 30 percent of needs (with wide variations across countries). AIDS remains the leading cause of death in Sub-Saharan Africa (1.6 million deaths in 2007). There are 300 to 500 million cases of malaria each year, leading to more than 1 million deaths. Nearly all the cases and more than 95 percent of the deaths occur in Sub-Saharan Africa.

7. **Ensure Environmental Sustainability**: Deforestation remains a critical problem, particularly in regions of biological diversity, which continues to decline. Greenhouse gas emissions are increasing faster than energy technology advancement.

8. **Develop a Global Partnership for Development**: Donor countries have renewed their commitment. Donors have to fulfill their pledges to match the current rate of core program development. Emphasis is being placed on the Bank Group's collaboration with multilateral and local partners to quicken progress toward the MDGs' realization.

Leadership

The President of the Bank, currently Jim Yong Kim, is responsible for chairing the meetings of the Boards of Directors and for overall management of the Bank. Traditionally, the Bank President has always been a US citizen nominated by the United States, the largest shareholder in the bank. The nominee is subject to confirmation by the Board of Executive Directors, to serve for a five-year, renewable term. While most World Bank presidents have had banking experience, some have not.[20] [21]

The vice presidents of the Bank are its principal managers, in charge of regions, sectors, networks and functions. There two Executive Vice Presidents, three Senior Vice Presidents, and 24 Vice Presidents.[22]

The Boards of Directors consist of the World Bank Group President and 25 Executive Directors. The President is the presiding officer, and ordinarily has no vote except a deciding vote in case of an equal division. The Executive Directors as individuals cannot exercise any power nor commit or represent the Bank unless specifically authorized by the Boards to do so. With the term beginning 1 November 2010, the number of Executive Directors increased by one, to 25.[23]

List of Presidents

Name	Dates	Nationality	Background
Eugene Meyer	1946–1946	United States	Newspaper publisher
John J. McCloy	1947–1949	United States	Lawyer and US Assistant Secretary of War
Eugene R. Black, Sr.	1949–1963	United States	Bank executive with Chase and executive director with the World Bank
George Woods	1963–1968	United States	Bank executive with First Boston Corporation
Robert McNamara	1968–1981	United States	US Defense Secretary, business executive with Ford Motor Company
Alden W. Clausen	1981–1986	United States	Lawyer, bank executive with Bank of America
Barber Conable	1986–1991	United States	New York State Senator and US Congressman
Lewis T. Preston	1991–1995	United States	Bank executive with J.P. Morgan
Sir James Wolfensohn	1995–2005	United States / Australia[24]	Corporate lawyer and banker
Paul Wolfowitz	2005–2007	United States	Various cabinet and government positions; US Ambassador to Indonesia, US Deputy Secretary of Defense
Robert Zoellick	2007–2012	United States	Bank executive with Goldman Sachs, Deputy Secretary of State and US Trade Representative
Jim Yong Kim	2012–present	United States	Korean-American physician and anthropologist, co-founder of Partners in Health and 17th President of Dartmouth College.[25] Elected on 16 April 2012.

José Antonio Ocampo, Ngozi Okonjo-Iweala, and Jim Yong Kim were candidates for the 2012 election. It was announced on 16 April 2012, that Jim Yong Kim will succeed Robert Zoellick as president, continuing the chain of successive World Bank president nominees from the United States. His nomination has been strongly opposed by the Nigerian Finance Minister Ngozi Okonjo-Iweala.[26]

List of chief economists

- Hollis B. Chenery (1972–1982)
- Anne Osborn Krueger (1982–1986)
- Stanley Fischer (1988–1990)
- Lawrence Summers (1991–1993)
- Michael Bruno (1993–1996)
- Joseph E. Stiglitz (1997–2000)
- Nicholas Stern (2000–2003)
- François Bourguignon (2003–2007)
- Justin Yifu Lin (June 2008–)

Justin Yifu Lin

Members

The International Bank for Reconstruction and Development (IBRD) has 187 member countries, while the International Development Association (IDA) has 171 members.[27] Each member state of IBRD should be also a member of the International Monetary Fund (IMF) and only members of IBRD are allowed to join other institutions within the Bank (such as IDA).[28]

Voting power

In 2010, voting powers at the World Bank were revised to increase the voice of developing countries, notably China. The countries with most voting power are now the United States (15.85%), Japan (6.84%), China (4.42%), Germany (4.00%), the United Kingdom (3.75%), France (3.75%), India (2.91%), Russia (2.77%), Saudi Arabia (2.77%) and Italy (2.64%). Under the changes, known as 'Voice Reform − Phase 2', countries other than China that saw significant gains included South Korea, Turkey, Mexico, Singapore, Greece, Brazil, India, and Spain. Most developed countries' voting power was reduced, along with a few poor countries such as Nigeria. The voting powers of the United States, Russia and Saudi Arabia were unchanged.[29] [30]

The changes were brought about with the goal of making voting more universal in regards to standards, rule-based with objective indicators, and transparent among other things. Now, developing countries have an increased voice in the "Pool Model," backed especially by Europe. Additionally, voting power is based on economic size in addition to International Development Association contributions.[31]

Poverty reduction strategies

For the poorest developing countries in the world, the bank's assistance plans are based on poverty reduction strategies; by combining a cross-section of local groups with an extensive analysis of the country's financial and economic situation the World Bank develops a strategy pertaining uniquely to the country in question. The government then identifies the country's priorities and targets for the reduction of poverty, and the World Bank aligns its aid efforts correspondingly.

Forty-five countries pledged US$25.1 billion in "aid for the world's poorest countries", aid that goes to the World Bank International Development Association (IDA) which distributes the loans to eighty poorer countries. While wealthier nations sometimes fund their own aid projects, including those for diseases, and although IDA is the recipient of criticism, Robert B. Zoellick, the president of the World Bank, said when the loans were announced on 15 December 2007, that IDA money "is the core funding that the poorest developing countries rely on".[32]

Clean Technology Fund management

The World Bank has been assigned temporary management responsibility of the Clean Technology Fund (CTF), focused on making renewable energy cost-competitive with coal-fired power as quickly as possible, but this may not continue after UN's Copenhagen climate change conference in December, 2009, because of the Bank's continued investment in coal-fired power plants.[33]

Training wings

World Bank Institute

The World Bank Institute [34] (WBI) creates learning opportunities for countries, World Bank staff and clients, and people committed to poverty reduction and sustainable development. WBI's work program includes training, policy consultations, and the creation and support of knowledge networks related to international economic and social development.

Global Development Learning Network

The Global Development Learning Network [35] (GDLN) is a partnership of over 120 learning centers (GDLN Affiliates) in nearly 80 countries around the world. GDLN Affiliates collaborate in holding events that connect people across countries and regions for learning and dialogue on development issues.

GDLN clients are typically NGOs, government, private sector and development agencies who find that they work better together on subregional, regional or global development issues using the facilities and tools offered by GDLN Affiliates. Clients also benefit from the ability of Affiliates to help them choose and apply these tools effectively, and to tap development practitioners and experts worldwide. GDLN Affiliates facilitate around 1000 videoconference-based activities a year on behalf of their clients, reaching some 90,000 people worldwide. Most of these activities bring together participants in two or more countries over a series of sessions. A majority of GDLN activities are organized by small government agencies and NGOs.

GDLN Asia Pacific

The GDLN in the East Asia and Pacific [36] region has experienced rapid growth and Distance Learning Centers now operate, or are planned in 20 countries: Australia, Mongolia, Cambodia, China, Indonesia, Singapore, Philippines, Sri Lanka, Japan, Papua New Guinea, South Korea, Thailand, Laos, Timor Leste, Fiji, Afghanistan, Bangladesh, India, Nepal and New Zealand. With over 180 Distance Learning Centers, it is the largest development learning network in the Asia and Pacific region. The Secretariat Office of GDLN Asia Pacific is located in the Center of Academic Resources of Chulalongkorn University, Bangkok, Thailand.

GDLN Asia Pacific was launched at the GDLN's East Asia and Pacific regional meeting held in Bangkok from 22 to 24 May 2006. Its vision is to become "the premier network exchanging ideas, experience and know-how across the Asia Pacific Region". GDLN Asia Pacific is a separate entity to The World Bank. It has endorsed its own Charter and Business Plan and, in accordance with the Charter, a GDLN Asia Pacific Governing Committee has been appointed.

The committee comprises China (2), Australia (1), Thailand (1), The World Bank (1) and finally, a nominee of the Government of Japan (1). The organization is currently hosted by Chulalongkorn University in Bangkok, Thailand, founding member of the GDLN Asia Pacific.

The Governing Committee has determined that the most appropriate legal status for the GDLN AP in Thailand is a "Foundation". The World Bank is currently engaging a solicitor in Thailand to process all documentation in order to obtain this legal status.

GDLN Asia Pacific is built on the principle of shared resources among partners engaged in a common task, and this is visible in the organizational structures that exist, as the network evolves. Physical space for its headquarters is provided by the host of the GDLN Centre in Thailand – Chulalongkorn University; Technical expertise and some infrastructure is provided by the Tokyo Development Learning Centre (TDLC); Fiduciary services are provided by Australian National University (ANU) Until the GDLN Asia Pacific is established as a legal entity tin Thailand, ANU, has offered to assist the governing committee, by providing a means of managing the inflow and outflow of funds and of reporting on them. This admittedly results in some complexity in contracting arrangements, which need to be worked out on a case by case basis and depends to some extent on the legal requirements of the countries involved.

Country assistance strategies

As a guideline to the World Bank's operations in any particular country, a Country Assistance Strategy is produced, in cooperation with the local government and any interested stakeholders and may rely on analytical work performed by the Bank or other parties.

Clean Air Initiative

Clean Air Initiative (CAI)[37] is a World Bank initiative to advance innovative ways to improve air quality in cities through partnerships in selected regions of the world by sharing knowledge and experiences. It includes electric vehicles.

United Nations Development Business

Based on an agreement between the United Nations and the World Bank in 1981, *Development Business* became the official source for World Bank Procurement Notices, Contract Awards, and Project Approvals.[38] In 1998, the agreement was re-negotiated, and included in this agreement was a joint venture to create an electronic version of the publication via the World Wide Web. Today, *Development Business* is the primary publication for all major multilateral development banks, United Nations agencies, and several national governments, many of whom have made the publication of their tenders and contracts in *Development Business* a mandatory requirement.[39] Currently, the subscription to "online version only" is not free, but costs US$ 550.[40]

The World Bank or the World Bank Group is also a sitting observer in the United Nations Development Group.[41]

Criticisms

The World Bank has long been criticized by non-governmental organizations, such as the indigenous rights group Survival International, and academics, including its former Chief Economist Joseph Stiglitz who is equally critical of the International Monetary Fund, the US Treasury Department, US and other developed country trade negotiators.[42] Critics argue that the so-called free market reform policies which the Bank advocates are often harmful to economic development if implemented badly, too quickly ("shock therapy"), in the wrong sequence or in weak, uncompetitive economies.[42] [43]

One of the strongest criticisms of the World Bank has been the way in which it is governed. While the World Bank represents 186 countries, it is run by a small number of economically powerful countries. These countries (which also provide most of the institution's funding) choose the leadership and senior management of the World Bank, and so their interests dominate the bank.[44]

In the 1990s, the World Bank and the IMF forged the Washington Consensus, policies which included deregulation and liberalization of markets, privatization and the downscaling of government. Though the Washington Consensus was conceived as a policy that would best promote development, it was criticized for ignoring equity, employment and how reforms like privatization were carried out. Joseph Stiglitz argued that the Washington Consensus placed too much emphasis on the growth of GDP, and not enough on the permanence of growth or on whether growth contributed to better living standards.[45]

Some analysis shows that the World Bank has increased poverty and been detrimental to the environment, public health and cultural diversity.[46]

Criticism of the World Bank often takes the form of protesting as seen in recent events such as the World Bank Oslo 2002 Protests,[47] the October Rebellion,[48] and the Battle of Seattle.[49] Such demonstrations have occurred all over the world, even amongst the Brazilian Kayapo people.[50]

Another source of criticism has been the tradition of having an American head the bank, implemented because the United States provides the majority of World Bank funding. "When economists from the World Bank visit poor countries to dispense cash and advice," observed *The Economist*, as yet another American, Jim Yong Kim, was lined up in 2012, "they routinely tell governments to reject cronyism and fill each important job with the best candidate available. It is good advice. The World Bank should take it."[51] Jim Yong Kim was duly named as president anyway.[52]

Structural adjustment

The effect of structural adjustment policies on poor countries has been one of the most significant criticisms of the World Bank. The 1979 energy crisis plunged many countries into economic crises.[53] The World Bank responded with structural adjustment loans which distributed aid to struggling countries while enforcing policy changes in order to reduce inflation and fiscal imbalance. Some of these policies included encouraging production, investment and labour-intensive manufacturing, changing real exchange rates and altering the distribution of government resources.[54] Structural adjustment policies were most effective in countries with an institutional framework that allowed these policies to be implemented easily.[55] For some countries, particularly in Sub-Saharan Africa, economic growth regressed and inflation worsened.[56] The alleviation of poverty was not a goal of structural adjustment loans, and the circumstances of the poor often worsened, due to a reduction in social spending and an increase in the price of food, as subsidies were lifted.[57]

By the late 1980s, international organizations began to admit that structural adjustment policies were worsening life for the world's poor. The World Bank changed structural adjustment loans, allowing for social spending to be maintained, and encouraging a slower change to policies such as transfer of subsidies and price rises.[58] In 1999, the World Bank and the IMF introduced the Poverty Reduction Strategy Paper approach to replace structural adjustment loans.[59] The Poverty Reduction Strategy Paper approach has been interpreted as an extension of structural adjustment policies as it continues to reinforce and legitimize global inequities.[60] Neither approach has addressed the inherent flaws within the global economy that contribute to economic and social inequities within developing countries.[61] By reinforcing the relationship between lending and client states, many believe that the World Bank has usurped indebted countries' power to determine their own economic policy.[62]

Sovereign immunity

Despite claiming goals of "good governance and anti-corruption"[63] the World Bank requires sovereign immunity from countries it deals with.[64] [65] [66] [67] [68] Sovereign immunity waives a holder from all legal liability for their actions. It is proposed that this immunity from responsibility is a "shield which [The World Bank] wants resort to, for escaping accountability and security by the people."[64] As the United States has veto power, it can prevent the World Bank from taking action against its interests.[64]

References

[1] http://web.worldbank.org/WBSITE/EXTERNAL/EXTABOUTUS/ORGANIZATION/BODEXT/
     0,,contentMDK:22427666~pagePK:64020054~piPK:64020408~theSitePK:278036,00.html

[2] "Board of Directors" (http://go.worldbank.org/11PWB3RTM0). World Bank. . Retrieved 14 August 2011.

[3] http://www.worldbank.org/

[4] "About Us" (http://web.worldbank.org/WBSITE/EXTERNAL/EXTABOUTUS/0,,pagePK:50004410~piPK:36602~theSitePK:29708,00.
     html). World Bank. 14 October 2008. . Retrieved 9 November 2008.

[5] "About Us" (http://web.worldbank.org/WBSITE/EXTERNAL/EXTABOUTUS/
     0,,contentMDK:20049563~pagePK:43912~menuPK:58863~piPK:36602,00.html#I1). World Bank. 29 June 2011. . Retrieved 14 August
     2011.

[6] "About The World Bank (FAQs)" (http://go.worldbank.org/1M3PFQQMD0). World Bank. April 2011. . Retrieved 14 August 2011.

[7] "The Founding Fathers" from the *IMF Archives* (http://jolis.worldbankimflib.org/Bwf/60panel3.htm)

[8] Goldman, Michael (2005). *Imperial Nature: The World Bank and Struggles for Social Justice in the Age of Globalization*. Yale University
     Press. pp. 52–54. ISBN 978-0-300-11974-9.

[9] Goldman, pp. 56–60.

[10] Bird, Kai (1992). *The Chairman: John J. McCloy, the Making of the American Establishment*. Simon and Schuster. pp. 288, 290–291. ISBN 978-0-671-45415-9.

[11] Goldman, pp. 60–63.

[12] Goldman, p. 62.

[13] Rotberg, Eugene. "Financial Operations of the World Bank." In *Bretton Woods: Looking to the Future*. ed. Bretton Woods Commission. Washington, D.C.: Bretton Woods Commission, 1994

[14] Mosley, Paul, Jane Harrigan, and John Toye. *Aid and Power: The World Bank and Policy-Based Lending*. London: Routledge, 1991

[15] Toussaint, Eric. *Your Money or Your Life! The Tyranny of Global Finance*. Pluto Press, 1998

[16] "About World Bank Administrative Tribunal" (http://lnweb90.worldbank.org/crn/wbt/wbtwebsite.nsf/(resultsweb)/ about?opendocument). *World Bank*. . Retrieved 14 August 2011.

[17] Cornia, Giovanni Andrea. *Adjustment with a Human Face*. 2 vols. Oxford: Clarendon Press, 1987–1988

[18] Goldman, pp. 93–97.

[19] Office of the Press Secretary, The White House (23 March 2012). "President Obama Announces U.S. Nomination of Dr. Jim Yong Kim to Lead World Bank" (http://www.whitehouse.gov/the-press-office/2012/03/23/ president-obama-announces-us-nomination-dr-jim-yong-kim-lead-world-bank). . Retrieved 23 March 2012.

[20] Hurlburt, Heather, "Why Jim Yong Kim would make a great World Bank president" (http://www.guardian.co.uk/commentisfree/ cifamerica/2012/mar/23/jim-yong-kim-world-bank-president), *The Guardian*, 23 Mar 2012.

[21] "Organization" (http://web.worldbank.org/WBSITE/EXTERNAL/EXTABOUTUS/ 0,,contentMDK:20040580~menuPK:1696997~pagePK:51123644~piPK:329829~theSitePK:29708,00.html). The World Bank Group. . Retrieved 25 March 2009.

[22] "Senior Management" (http://web.worldbank.org/WBSITE/EXTERNAL/EXTABOUTUS/ORGANIZATION/EXTPRESIDENT2007/ 0,,contentMDK:20040913~menuPK:64822291~pagePK:64821878~piPK:64821912~theSitePK:3916065,00.html). The World Bank Group. 2012-03-21. . Retrieved 2012-04-17.

[23] "Boards of Directors" (http://web.worldbank.org/WBSITE/EXTERNAL/EXTABOUTUS/ 0,,contentMDK:22494475~menuPK:8336906~pagePK:51123644~piPK:329829~theSitePK:29708,00.html). The World Bank Group. 2012-01-19. . Retrieved 2012-04-17.

[24] The World Bank President is traditionally an American citizen. Wolfensohn was a naturalised American citizen before taking office.

[25] http://ibnlive.in.com/news/jim-yong-kim-chosen-to-head-world-bank/249266-2.html

[26] *BBC News*. 16 April 2012. http://www.bbc.co.uk/news/business-17735368. Retrieved 16 April 2012.

[27] "Members" (http://go.worldbank.org/Y33OQYNE90). The World Bank Group. . Retrieved 6 February 2008.

[28] "Member countries" (http://go.worldbank.org/CFDWBOOR50). The World Bank Group. 19 January 2012. . Retrieved 19 February 2012.

[29] "IBRD 2010 Voting Power Realignment" (http://siteresources.worldbank.org/NEWS/Resources/ IBRD2010VotingPowerRealignmentFINAL.pdf) (PDF). Siteresources.worldbank.org. . Retrieved 14 August 2011. "Source: World Bank Group Voice Reform: Enhancing Voice and Participation in Developing and Transition Countries in 2010 and Beyond, DC 2010-0006/1, April 25, 2010"

[30] China given more influence in World Bank (http://www.rthk.org.hk/rthk/news/englishnews/news.htm?main&20100426&56& 663699), RTHK, 26 April 2010

[31] Stumm, Mario (March 2011). "World Bank: More responsibility for developing countries" (http://www.inwent.org/ez/articles/193054/ index.en.shtml). Inwent.org. . Retrieved 12 August 2011.

[32] Landler, Mark (15 December 2007). "Britain Overtakes U.S. as Top World Bank Donor" (http://www.nytimes.com/2007/12/15/world/ 15worldbank.html). *The New York Times*. . Retrieved 14 August 2011.

[33] "Global Development: Views from the Center" (http://blogs.cgdev.org/globaldevelopment/2008/05/climate_change_in_nashville_a. php). Center for Global Development. 20 May 2008. . Retrieved 9 November 2008.

[34] http://www.worldbank.org/wbi/home.html

[35] http://www.gdln.org

[36] http://www.gdlnap.org/

[37] "CAI Global" (http://www.cleanairnet.org). Cleanairnet.org. . Retrieved 31 May 2010.

[38] "UN Development Business Subscriber Log In page" (http://www.devbusiness.com). Devbusiness.com. . Retrieved 14 August 2011.

[39] "About UN Development Business" (http://www.devbusiness.com/about.asp). Devbusiness.com. . Retrieved 14 August 2011.

[40] "Subscribe to UN Development Business Online" (http://www.devbusiness.com/subscriptioninformation.asp). Devbusiness.com. . Retrieved 14 August 2011.

[41] "UNDG Members" (http://www.undg.org/index.cfm?P=13). Undg.org. . Retrieved 12 August 2011.

[42] See Joseph Stiglitz, *The Roaring Nineties, Globalization and Its Discontents*, and *Making Globalization Work*.

[43] MacClancy, Jeremy (2002). *Exotic No More: Anthropology on the Front Lines*. University Of Chicago Press. ISBN 0-226-50013-6.

[44] Woods, Ngaire. *The Globalizers: The IMF, the World Bank, and Their Borrowers*. Ithica and London: Cornell University Press, 2006, pp.190

[45] Stiglitz, Joseph E. *Making Globalization Work*. New York and London: W.W. Norton & Company, 2006, pp. 17

[46] "Criticism of World Trade Organization, World Bank and International Monetary Fund – Editorial" (http://findarticles.com/p/articles/ mi_m2465/is_6_30/ai_65653637). The Ecologist (original), later republished at BNET Business Network. 2000-09. . Retrieved 7 October

2007.

[47] Gibbs, Walter (25 June 2002). "World Briefing – Europe: Norway: Protests As World Bank Meets" (http://www.nytimes.com/2002/06/ 25/world/world-briefing-europe-norway-protests-as-world-bank-meets.html?n=Top/Reference/Times Topics/Subjects/F/Foreign Aid). *New York Times.* .

[48] "Violence Erupts at Protest in Georgetown" (http://www.washingtonpost.com/wp-dyn/content/article/2007/10/19/ AR2007101901728.html). *The Washington Post*: p. B01. 20 October 2007. . Retrieved 30 May 2008.

[49] Kimberly A.C. Wilson, Embattled police chief resigns (http://www.seattlepi.com/local/cops071.shtml), *Seattle Post-Intelligencer*, 7 December 1999. Accessed online 19 May 2008.

[50] Clendenning, Alan (Altamira, Brazil) (21 May 2008). "Amazon Indians Attack Official Over Dam Project" (http://news. nationalgeographic.com/news/2008/05/080521-AP-indians-dam.html). Associated Press. .

[51] "Hats off to Ngozi" (http://www.economist.com/node/21551490). *The Economist*. 31 March 2012. . Retrieved 2 April 2012.

[52] Dominic Rushe; Heather Stewart; Monica Mark (16 April 2012). "World Bank names US-nominated Jim Yong Kim as president" (http:// www.guardian.co.uk/business/2012/apr/16/world-bank-president-jim-yong-kim). guardian.co.uk. . Retrieved 17 April 2012.

[53] deVries, Barend A. (1996). "The World Bank's Focus on Poverty". In Griesgraber, Jo Marie; Gunter, Bernhard G.. *The World Bank: Lending on a Global Scale*. Pluto Press. p. 68. ISBN 978-0-7453-1049-7.

[54] deVries, p. 69.

[55] deVries, p. 69.

[56] deVries, p. 69.

[57] deVries, p. 69.

[58] deVries, p. 70.

[59] Tan, Celine (2007). "The poverty of amnesia: PRSPs in the legacy of structural adjustment". In Stone, Diane; Wright, Christopher. *The World Bank and Governance: A Decade of Reform and Reaction*. Routledge. p. 147. ISBN 978-0-415-41282-7.

[60] Tan, p. 152.

[61] Tan, p. 152.

[62] Chossudovsky M. *The Globalization of Poverty: Impacts of IMF and World Bank Reforms*. Penang: Third World Network, 1997 in Tan, 152

[63] "Fraud and Corruption" (http://web.worldbank.org/WBSITE/EXTERNAL/EXTABOUTUS/ORGANIZATION/ORGUNITS/ EXTETHICS/0,,contentMDK:20835544~menuPK:2273501~pagePK:64168445~piPK:64168309~theSitePK:593304,00.html). World Bank. 24 October 2009. .

[64] "The World Bank and the Question of Immunity" (http://www.unnayan.org/Other/IFI_Watch_Bangladesh_Vol_1 No_1.pdf). IFI Watch Bangladesh. 4 September 2004. .

[65] "Sovereign Immunity" (http://siteresources.worldbank.org/INTINFANDLAW/Resources/sovereignimmunity.pdf). World Bank. . Retrieved 24 October 2009.

[66] Adam Isaac Hasson. "Extraterritorial jurisdiction and sovereign immunity on trial: Noriega, Pinochet and Milosevic – Trends in political accountability and transnational criminal law" (http://lawdigitalcommons.bc.edu/iclr/vol25/iss1/6/). Archived from the original (http:// www.bc.edu/bc_org/avp/law/lwsch/journals/bciclr/25_1/05_TXT.htm) on 2012-04-25. . Retrieved 23 October 2009.

[67] "Crime and Reward: Immunity To The World Bank" (http://www.countercurrents.org/gl-muhammad061104.htm). 6 November 2004. .

[68] "Water Policies and the International Financial Institutions" (http://www.citizenarchive.org/cmep/Water/cmep_Water/wbimf/index. cfm?relatedpages=1&catID=106&secID=1592). Public Citizen. Archived from the original (http://www.citizen.org/cmep/Water/ cmep_Water/wbimf/) on 2012-04-25. .

Notes

External links

- Official website (http://www.worldbank.org/)
- IBRD main page (http://web.worldbank.org/WBSITE/EXTERNAL/EXTABOUTUS/EXTIBRD/ 0,,menuPK:3046081~pagePK:64168427~piPK:64168435~theSitePK:3046012,00.html)
- IDA main page (http://www.worldbank.org/ida/)
- Access to Information page (http://web.worldbank.org/WBSITE/EXTERNAL/ PROJECTANDOPERATIONS/EXTINFODISCLOSURE/ 0,,menuPK:64864911~pagePK:4749265~piPK:4749256~theSitePK:5033734,00.html)

Digital_divide

The **Digital Divide** refers to any inequalities between groups, broadly construed, in terms of access to, use of, or knowledge of information and communication technologies (ICT).[1] [2] The divide inside countries (such as the digital divide in the United States) can refer to inequalities between individuals, households, businesses, and geographic areas at different socioeconomic and other demographic levels, while[3] [4] [5] the Global digital divide designates countries as the units of analysis and examines the divide between developing and developed countries on an international scale.[2]

Approaches

Conceptualization of the digital divide is often as follows:[6] [7]

1. Theoretical explanations for the digital divide, or who connects with which attributes: demographic characteristics of connected individuals and their cohorts.
2. Means of connectivity, or how individuals and their cohorts are connecting and to what: infrastructure, location, and network availability.
3. Intensity of connectivity, or how sophisticated the usage: mere access, retrieval, interactivity, innovative contributions.
4. Purpose of connectivity, or why individuals and their cohorts are connecting: reasons individuals are online and uses of the Internet and ICTs.
5. Lack of connectivity, or why individuals and their cohorts are not connecting.

In coveted research, while each explanation is examined, the others should be controlled for in order to eliminate interaction effects or mediating variables,[8] but these explanations are meant to stand as general trends, not direct causes. Each of the above listed items can be looked at from different angles, which leads to a myriad of ways to look at (or define) the digital divide. For example, measurements for the intensity of usage, such as incidence and frequency, vary by study. Some report usage as access to Internet and ICTs while others report usage as having previously connected to the Internet. Some studies focus on specific technologies, others on a combination (such as Infostate, proposed by Orbicom-UNESCO, the Digital Opportunity Index, or ITU's ICT Development Index). Based on these differences, there are hundreds of alternatives ways to define the digital divide.[7]

Explanatory variables

Research suggests a multitude of explanations for the digital divide including, but not limited to: education, income[9] , age, skills, awareness, race, ethnic origin, location, and gender (which is contested as being a confounding variable[10]); political and cultural access; and psychological attitudes to Internet access and usage.[11] [8] [12] [13] [14] [15] [16] Income levels and educational attainment are identified as having the largest explanatory power to explain ICT access and usage, with age being a third important variable[7] [9] , which means that prototypical "victims" to the digital divide can be foremost characterized as poorer, less educated, and older.

Means of connectivity

Infrastructure

The infrastructure by which individuals, households, businesses, and communities connect to the Internet address the physical mediums that people use to connect to the Internet such as desktop computers, laptops, cell phones, iPods or other MP3 players, Xboxes or Play Stations, electronic books readers, and tablets such as iPads.[17]

Location

Internet connectivity can be utilized at a variety of locations such as homes, offices, schools, libraries, public spaces, Internet cafes, etc. There are also varying levels of connectivity in rural, suburban, and urban areas.[18]

Overcoming the digital divide

An individual must be able to connect in order to achieve enhancement of social and cultural capital and achieve mass economic gains in productivity. Therefore access is a necessary (but not sufficient) condition for overcoming the digital divide. Access to ICT meets significant challenges that stem from income restrictions. The borderline between ICT as a necessity good and ICT as a luxury good is roughly around the "magical number" of US$10 per person per month, or US$120 per year[9] , which means that people consider ICT expenditure of US$120 per year as a basic necessity. Since more than 40% of the world population lives on less than US$ 2 per day, and around 20% live on less than US$ 1 per day (or less than US$ 365 per year), these income segments would have to spend one third of their income on ICT (120/365 = 33%), which is a lot to ask, since the global average of ICT spending is at a mere 3% of income.[9] Potential solutions include driving down the costs or ICT, which includes low cost technologies and shared access through Telecentres.

Furthermore, even though individuals might be capable of accessing the Internet, many are thwarted by barriers to entry such as a lack of means to infrastructure or the inability to comprehend the information that the Internet provides. Lack of adequate infrastructure and lack of knowledge are two major obstacles that impede mass connectivity. These barriers limit individuals' capabilities in what they can do and what they can achieve in accessing technology. Some individuals have the ability to connect, but have nonfunctioning capabilities in that they do not have the knowledge to use what information ICTs and Internet technologies provide them. This leads to a focus on capabilities and skills, as well as awareness to move from mere access to effective usage of ICT[19] .

The United Nations is aiming to raise awareness of the divide by way of the World Information Society Day which has taken place yearly since May 17, 2001.[20] It also set up the Information and Communications Technology (ICT) Task Force in November 2001.[21]

Implications

Social capital

Once an individual is connected, Internet connectivity and ICTs can enhance their future social and cultural capital. Social capital is acquired through repeated interactions with other individuals or groups of individuals. Connecting to the Internet creates another set of means by which to achieve repeated interactions. ICTs and Internet connectivity enable repeated interactions through access to social networks, chat rooms, and gaming sites. Once an individual has access to connectivity, obtains infrastructure by which to connect, and can understand and use the information that ICTs and connectivity provide, that individual is capable of becoming a "digital citizen".[8]

Criticisms

Second-level digital divide

The second-level digital divide, also referred to as the production gap, describes the gap that separates the consumers of content on the internet from the producers of content.[22] As the technological digital divide is decreasing between those with access to the internet and those without, the meaning of the term digital divide is evolving.[23] Previously, digital divide research has focused on accessibility to the internet and internet consumption. However, with more and more of the population with access to the internet, researchers are examining how people use the internet to create content and what impact socioeconomics are having on user behavior.[24] New applications have made it possible for anyone with a computer and an internet connection to be a creator of content, yet the majority of user generated content available widely on the internet, like public blogs, is created by a small portion of the internet using population. Web 2.0 technologies like Facebook, YouTube, Twitter, and Blogs enable users to participate online and create content without having to understand how the technology actually works, leading to an ever increasing digital divide between those who have the skills and understanding to interact more fully with the technology and those who are passive consumers of it.[22] Many are only nominal content creators through the use of Web 2.0, like posting photos and status updates on Facebook, but not truly interacting with the technology. Some of the reasons for this production gap include material factors like what type of internet connection one has and the frequency of access to the internet. The more frequently a person has access to the internet and the faster the connection, the more opportunities they have to gain the technology skills and the more time they have to be creative.[25] Other reasons include cultural factors often associated with class and socioeconomic status. Users of lower socioeconomic status are less likely to participate in content creation due to disadvantages in education and lack of the necessary free time for the work involved in blog or web site creation and maintenance.[25]

The knowledge divide

Since gender, age, racial, income, and educational gaps in the digital divide have lessened compared to past levels, some researchers suggest that the digital divide is shifting from a gap in access and connectivity to ICTs to a knowledge divide.[23] A knowledge divide concerning technology presents the possibility that the gap has moved beyond access and having the resources to connect to ICTs to interpreting and understanding information presented once connected.[26]

See also

- Achievement gap
- Computer technology for developing areas
- Digital Society Day, each 17 October in India
- Digital Opportunity Index
- Generation gap
- Income gap
- Information society
- Knowledge divide
- National broadband plans from around the world
- Rural Internet
- Groups devoted to digital divide issues:
    - United Nations Information and Communication Technologies Task Force
    - Center for Digital Inclusion
    - Close the Gap International VZW

- Digital Textbook a South Korean Project that intends to distribute tablet notebooks to elementary school students.
- Low cost notebooks and subnotebooks
  - Classmate PC notebook by Intel
  - Eee PC subnotebook by ASUS
  - Sinomanic subnotebook (250$)
  - XO Laptop subnotebook for developing regions,One Laptop per Child (OLPC)
  - Zonbu
  - VIA pc-1 Initiative, VIA Technologies digital divide program
- Web 2.0

References

[1] U.S. Department of Commerce, National Telecommunications and Information Administration (NTIA). 1995. Falling through the net: A survey of the "have nots" in rural and urban America. Retrieved from http://www.ntia.doc.gov/ntiahome/fallingthru.html.

[2] Chinn, Menzie D. and Robert W. Fairlie. 2004. The Determinants of the Global Digital Divide: A Cross-Country Analysis of Computer and Internet Penetration. Economic Growth Center. Retrieved from http://www.econ.yale.edu/growth_pdf/cdp881.pdf.

[3] Norris, P. 2001. Digital divide: Civic engagement, information poverty and the Internet world-wide. Cambridge, MA: Cambridge Univ. Press.

[4] U.S. Department of Commerce, National Telecommunications and Information Administration (NTIA). 1995. Falling through the net: A survey of the "have nots" in rural and urban America. Retrieved from http://www.ntia.doc.gov/ntiahome/fallingthru.html.

[5] Patricia, J.P. 2003. E-government, E-Asean Task force, UNDP-APDIP. From: http://www.apdip.net/publications/iespprimers/eprimer-egov.pdf

[6] Buente, Wayne, and Alice Robbin. 2008. Trends in Internet Information Behavior, 2000-2004. Journal of the American Society for Information Science and Technology 59(11): 1743-1760. www.interscience.wiley.com

[7] Hilbert, M. 2011. The end justifies the definition: The manifold outlooks on the digital divide and their practical usefulness for policy-making (http://dx.doi.org/10.1016/j.telpol.2011.06.012). Telecommunications Policy, 35(8), 715-736. Retrieved from: http://martinhilbert.net/ManifoldDigitalDivide_Hilbert_AAM.pdf

[8] Mossberger, Karen, Carolina J. Tolbert, and Michele Gilbert. 2006. Race, Place, and Information Technology (IT). Urban Affairs Review. 41:583-620. retrieved from http://uar.sagepub.com/content/41/5/583

[9] Martin Hilbert *"When is Cheap, Cheap Enough to Bridge the Digital Divide? Modeling Income Related Structural Challenges of Technology Diffusion in Latin America"* (http://dx.doi.org/10.1016/j.worlddev.2009.11.019). World Development (http://www.elsevier.com/locate/worlddev), Volume 38, issue 5, p. 756-770; free access to the study here: martinhilbert.net/CheapEnoughWD_Hilbert_pre-print.pdf

[10] "Digital gender divide or technologically empowered women in developing countries? A typical case of lies, damned lies, and statistics " (http://dx.doi.org/10.1016/j.wsif.2011.07.001), Martin Hilbert (2011), Women's Studies International Forum, 34(6), 479-489; free access to the study here: martinhilbert.net/DigitalGenderDivide.pdf

[11] Lawton, Tait. "15 Years of Chinese Internet Usage in 13 Pretty Graphs" (http://www.east-west-connect.com/15-years-chinese-internet-usage-13-pretty-graphs). *East-West-Connect.com*. CNNIC. .

[12] Wang, Wensheng. Impact of ICTs on Farm Households in China, ZEF of University Bonn, 2001

[13] http://www.cnnic.net.cn/uploadfiles/pdf/2007/2/14/200607.pdf Statistical Survey Report on the Internet Development in China. Jan, 2007.

[14] Guillen, M. F., & Suárez, S. L. (2005). Explaining the global digital divide: Economic, political and sociological drivers of cross-national internet use. Social Forces, 84(2), 681-708.

[15] Wilson, III. E.J. (2004). The Information Revolution and Developing Countries. Cambridge, MA: The MIT Press.

[16] Carr, Deborah (2007). The Global Digital Divide. Contexts, 6(3), 58-58. Retrieved from http://ezproxy.qa.proquest.com/docview/219574259?accountid=14771.

[17] Zickuher, Kathryn. 2011. Generations and their gadgets. Pew Internet & American Life Project. Retrieved from: http://www.pewinternet.org/Reports/2011/Generations-and-gadgets/Report/Desktop-and-Laptop-Computers.aspx.

[18] Livingston, Gretchen. 2010. Latinos and Digital Technology, 2010. Pew Hispanic Center

[19] Karen Mossberger (2003). Virtual Inequality: Beyond the Digital Divide. Georgetown University Press

[20] United Nations Educational (http://www.unesco.org/new/en/communication-and-information/resources/news-and-in-focus-articles/in-focus-articles/) UNDay

[21] http://blog.unicttaskforce.org/unicttaskforce

[22] Reilley, Collen A. Teaching Wikipedia as a Mirrored Technology. First Monday, Vol. 16, No. 1-3, January 2011

[23] Graham, Mark. (2011). Time Machines and Virtual Portals: The Spatialities of the Digital Divide. Progress in Development Studies, Vol. 11, No. 3, 211-227. Retrieved from: http://pdj.sagepub.com/content/11/3/211

[24] Correa, Teresa. (2008) Literature Review: Understanding the "second-level digital divide" papers by Teresa Correa. Unpublished
       manuscript, School of Journalism, College of Communication, University of Texas at Austin. (http://utexas.academia.edu/TeresaCorrea/
       Papers/140182/Literature_Review_Understading_the_second-level_digital_divide_).
[25] (http://sociology.berkeley.edu/documents/newspage_docs/J.Schradie - The Digital Production Gap for Web Post Poetics.pdf).
       Schradie, Jen. The Digital Production Gap: The Digital Divide and Web 2.0 Collide. Poetics, Vol. 39, No. 2. April 2011, p. 145-168.
[26] Sciadas, George. (2003). Monitoring the Digital Divide…and Beyond. Orbicom.

External links

- E-inclusion (http://ec.europa.eu/information_society/activities/einclusion/index_en.htm)
- An information society for all (http://ec.europa.eu/information_society/tl/soccul/eincl/index_en.htm)
- Digital Inclusion Network (http://forums.e-democracy.org/groups/inclusion)

Green_computing

Green computing, **green IT** or **ICT Sustainability**, refers to environmentally sustainable computing or IT. In the article *Harnessing Green IT: Principles and Practices*, San Murugesan defines the field of green computing as "the study and practice of designing, manufacturing, using, and disposing of computers, servers, and associated subsystems—such as monitors, printers, storage devices, and networking and communications systems — efficiently and effectively with minimal or no impact on the environment."[1] The goals of green computing are similar to green chemistry; reduce the use of hazardous materials, maximize energy efficiency during the product's lifetime, and promote the recyclability or biodegradability of defunct products and factory waste. Research continues into key areas such as making the use of computers as energy-efficient as possible, and designing algorithms and systems for efficiency-related computer technologies.

Origins

In 1992, the U.S. Environmental Protection Agency launched Energy Star, a voluntary labeling program that is designed to promote and recognize energy-efficiency in monitors, climate control equipment, and other technologies. This resulted in the widespread adoption of sleep mode among consumer electronics. Concurrently, the Swedish organization TCO Development launched the TCO Certification program to promote low magnetic and electrical emissions from CRT-based computer displays; this program was later expanded to include criteria on energy consumption, ergonomics, and the use of hazardous materials in construction.[2]

Energy Star logo

Regulations and industry initiatives

The Organisation for Economic Co-operation and Development (OECD) has published a survey of over 90 government and industry initiatives on "Green ICTs", i.e. information and communication technologies, the environment and climate change. The report concludes that initiatives tend to concentrate on the greening ICTs themselves rather than on their actual implementation to tackle global warming and environmental degradation. In general, only 20% of initiatives have measurable targets, with government programs tending to include targets more frequently than business associations.[3]

Government

Many governmental agencies have continued to implement standards and regulations that encourage green computing. The Energy Star program was revised in October 2006 to include stricter efficiency requirements for computer equipment, along with a tiered ranking system for approved products.[4] [5]

Some efforts place responsibility on the manufacturer to dispose of the equipment themselves after it is no longer needed; this is called the extended producer responsibility model. The European Union's directives 2002/95/EC (Restriction of Hazardous Substances Directive), on the reduction of hazardous substances, and 2002/96/EC (Waste Electrical and Electronic Equipment Directive) on waste electrical and electronic equipment required the substitution of heavy metals and flame retardants like Polybrominated biphenyl and Polybrominated diphenyl ethers in all electronic equipment put on the market starting on July 1, 2006. The directives placed responsibility on manufacturers for the gathering and recycling of old equipment.[6]

There are currently 26 US states that have established state-wide recycling programs for obsolete computers and consumer electronics equipment.[7] The statutes either impose an "advance recovery fee" for each unit sold at retail or require the manufacturers to reclaim the equipment at disposal.

In 2010, the American Recovery and Reinvestment Act (ARRA) was signed into legislation by President Obama. The bill allocated over $90 billion to be invested in green initiatives (renewable energy, smart grids, energy efficiency, etc.) In January 2010, the U.S. Energy Department granted $47 million of the ARRA money towards projects that aim to improve the energy efficiency of data centers. The projects will provide research on the following three areas: optimize data center hardware and software, improve power supply chain, and data center cooling technologies.[8]

Industry

- Climate Savers Computing Initiative (CSCI) is an effort to reduce the electric power consumption of PCs in active and inactive states.[9] The CSCI provides a catalog of green products from its member organizations, and information for reducing PC power consumption. It was started on 2007-06-12. The name stems from the World Wildlife Fund's Climate Savers program, which was launched in 1999.[10] The WWF is also a member of the Computing Initiative.[9]
- The Green Electronics Council offers the Electronic Product Environmental Assessment Tool (EPEAT) to assist in the purchase of "greener" computing systems. The Council evaluates computing equipment on 51 criteria - 23 required and 28 optional - that measure a product's efficiency and sustainability attributes. Products are rated Gold, Silver, or Bronze, depending on how many optional criteria they meet. On 2007-01-24, President George W. Bush issued Executive Order 13423, which requires all United States Federal agencies to use EPEAT when purchasing computer systems.[11] [12]
- The Green Grid is a global consortium dedicated to advancing energy efficiency in data centers and business computing ecosystems. It was founded in February 2007 by several key companies in the industry – AMD, APC, Dell, HP, IBM, Intel, Microsoft, Rackable Systems, SprayCool, Sun Microsystems and VMware. The Green Grid has since grown to hundreds of members, including end-users and government organizations, all focused on improving data center infrastructure efficiency (DCIE).
- The Green500 list rates supercomputers by energy efficiency (megaflops/watt, encouraging a focus on efficiency rather than absolute performance.
- Green Comm Challenge is an organization that promotes the development of energy conservation technology and practices in the field of Information and Communications Technology (ICT).
- The Transaction Processing Performance Council(TPC) Energy specification augments the existing TPC benchmarks by allowing for optional publications of energy metrics alongside their performance results.[13]
- The SPEC Power is the first industry standard benchmark that measures power consumption in relation to performance for server-class computers.

Approaches

In the article *Harnessing Green IT: Principles and Practices*, San Murugesan defines the field of green computing as "the study and practice of designing, manufacturing, using, and disposing of computers, servers, and associated subsystems — such as monitors, printers, storage devices, and networking and communications systems — efficiently and effectively with minimal or no impact on the environment."[1] Murugesan lays out four paths along which he believes the environmental effects of computing should be addressed:[1] Green use, green disposal, green design, and green manufacturing. Green computing can also develop solutions that offer benefits by "aligning all IT processes and practices with the core principles of sustainability, which are to reduce, reuse, and recycle; and finding innovative ways to use IT in business processes to deliver sustainability benefits across the enterprise and beyond". [14]

Modern IT systems rely upon a complicated mix of people, networks, and hardware; as such, a green computing initiative must cover all of these areas as well. A solution may also need to address end user satisfaction, management restructuring, regulatory compliance, and return on investment (ROI). There are also considerable fiscal motivations for companies to take control of their own power consumption; "of the power management tools available, one of the most powerful may still be simple, plain, common sense."[15]

Product longevity

Gartner maintains that the PC manufacturing process accounts for 70 % of the natural resources used in the life cycle of a PC.[16] More recently, Fujitsu released a Life Cycle Assessment (LCA) of a desktop that show that manufacturing and end of life accounts for the majority of this laptop ecological footprint.[17] Therefore, the biggest contribution to green computing usually is to prolong the equipment's lifetime. Another report from Gartner recommends to "Look for product longevity, including upgradability and modularity." [18] For instance, manufacturing a new PC makes a far bigger ecological footprint than manufacturing a new RAM module to upgrade an existing one.

Data center design

Data center facilities are heavy consumers of energy, accounting for between 1.1% and 1.5% of the world's total energy use in 2010 [1]. The U.S. Department of Energy estimates that data center facilities consume up to 100 to 200 times more energy than standard office buildings [19] .

Energy efficient data center design should address all of the energy use aspects included in a data center: from the IT equipment to the HVAC equipment to the actual location, configuration and construction of the building.

The U.S. Department of Energy specifies five primary areas on which to focus energy efficient data center design best practices [20] :

- Information technology (IT) systems
- Environmental conditions
- Air management
- Cooling systems
- Electrical systems

Additional energy efficient design opportunities specified by the U.S. Department of Energy include on-site electrical generation and recycling of waste heat [21] .

Energy efficient data center design should help to better utilize a data center's space, and increase performance and efficiency.

Software and deployment optimization

Algorithmic efficiency

The efficiency of algorithms has an impact on the amount of computer resources required for any given computing function and there are many efficiency trade-offs in writing programs. While algorithmic efficiency does not have as much impact as other approaches , it is still an important consideration. A study by a physicist at Harvard, estimated that the average Google search released 7 grams of carbon dioxide (CO_2).[22] However, Google disputes this figure, arguing instead that a typical search produces only 0.2 grams of CO_2.[23] More recently, an independent study by GreenIT.fr [24] demonstrate that Windows 7 + Office 2010 require 70 times more memory (RAM) than Windows 98 + Office 2000 to write exactly the same text or send exactly the same e-mail than 10 years ago[25]

Resource allocation

Algorithms can also be used to route data to data centers where electricity is less expensive. Researchers from MIT, Carnegie Mellon University, and Akamai have tested an energy allocation algorithm that successfully routes traffic to the location with the cheapest energy costs. The researchers project up to a 40 percent savings on energy costs if their proposed algorithm were to be deployed. However, this approach does not actually reduce the amount of energy being used; it reduces only the cost to the company using it. Nonetheless, a similar strategy could be used to direct traffic to rely on energy that is produced in a more environmentally friendly or efficient way. A similar approach has also been used to cut energy usage by routing traffic away from data centers experiencing warm weather; this allows computers to be shut down to avoid using air conditioning.[26]

Larger server centers are sometimes located where energy and land are inexpensive and readily available. Local availability of renewable energy, climate that allows outside air to be used for cooling, or locating them where the heat they produce may be used for other purposes could be factors in green siting decisions.

Virtualization

Computer virtualization refers to the abstraction of computer resources, such as the process of running two or more logical computer systems on one set of physical hardware. The concept originated with the IBM mainframe operating systems of the 1960s, but was commercialized for x86-compatible computers only in the 1990s. With virtualization, a system administrator could combine several physical systems into virtual machines on one single, powerful system, thereby unplugging the original hardware and reducing power and cooling consumption. Virtualization can assist in distributing work so that servers are either busy or put in a low-power sleep state. Several commercial companies and open-source projects now offer software packages to enable a transition to virtual computing. Intel Corporation and AMD have also built proprietary virtualization enhancements to the x86 instruction set into each of their CPU product lines, in order to facilitate virtualized computing.

Terminal servers

Terminal servers have also been used in green computing. When using the system, users at a terminal connect to a central server; all of the actual computing is done on the server, but the end user experiences the operating system on the terminal. These can be combined with thin clients, which use up to 1/8 the amount of energy of a normal workstation, resulting in a decrease of energy costs and consumption. There has been an increase in using terminal services with thin clients to create virtual labs. Examples of terminal server software include Terminal Services for Windows and the Linux Terminal Server Project (LTSP) for the Linux operating system.

Power management

The Advanced Configuration and Power Interface (ACPI), an open industry standard, allows an operating system to directly control the power-saving aspects of its underlying hardware. This allows a system to automatically turn off components such as monitors and hard drives after set periods of inactivity. In addition, a system may hibernate, where most components (including the CPU and the system RAM) are turned off. ACPI is a successor to an earlier Intel-Microsoft standard called Advanced Power Management, which allows a computer's BIOS to control power management functions.

Some programs allow the user to manually adjust the voltages supplied to the CPU, which reduces both the amount of heat produced and electricity consumed. This process is called undervolting. Some CPUs can automatically undervolt the processor, depending on the workload; this technology is called "SpeedStep" on Intel processors, "PowerNow!"/"Cool'n'Quiet" on AMD chips, LongHaul on VIA CPUs, and LongRun with Transmeta processors.

Data center power

Data centers, which have been criticized for its extraordinary high energy demand, are a primary focus for proponents of green computing.[27] The U.S. federal government has set a minimum 10% reduction target for data center energy usage by 2011.[27] With the aid of a self-styled ultraefficient evaporative cooling technology, Google Inc. has been able to reduce its energy consumption to 50% of that of the industry average.[27]

Operating system support

The dominant desktop operating system, Microsoft Windows, has included limited PC power management features since Windows 95.[28] These initially provided for stand-by (suspend-to-RAM) and a monitor low power state. Further iterations of Windows added hibernate (suspend-to-disk) and support for the ACPI standard. Windows 2000 was the first NT-based operating system to include power management. This required major changes to the underlying operating system architecture and a new hardware driver model. Windows 2000 also introduced Group Policy, a technology that allowed administrators to centrally configure most Windows features. However, power management was not one of those features. This is probably because the power management settings design relied upon a connected set of per-user and per-machine binary registry values,[29] effectively leaving it up to each user to configure their own power management settings.

This approach, which is not compatible with Windows Group Policy, was repeated in Windows XP. The reasons for this design decision by Microsoft are not known, and it has resulted in heavy criticism.[30] Microsoft significantly improved this in Windows Vista[31] by redesigning the power management system to allow basic configuration by Group Policy. The support offered is limited to a single per-computer policy. The most recent release, Windows 7 retains these limitations but does include refinements for more efficient user of operating system timers, processor power management,[32] [33] and display panel brightness. The most significant change in Windows 7 is in the user experience. The prominence of the default High Performance power plan has been reduced with the aim of encouraging users to save power.

There is a significant market in third-party PC power management software offering features beyond those present in the Windows operating system.[34] [35] [36] available. Most products offer Active Directory integration and per-user/per-machine settings with the more advanced offering multiple power plans, scheduled power plans, anti-insomnia features and enterprise power usage reporting. Notable vendors include 1E NightWatchman.,[37] [38] Data Synergy PowerMAN (Software),[39] Faronics Power Save[40] and Verdiem SURVEYOR.[41]

Power supply

Desktop computer power supplies (PSUs) are in general 70–75% efficient,[42] dissipating the remaining energy as heat. An industry initiative called 80 PLUS certifies PSUs that are at least 80% efficient; typically these models are drop-in replacements for older, less efficient PSUs of the same form factor.[43] As of July 20, 2007, all new Energy Star 4.0-certified desktop PSUs must be at least 80% efficient.[44]

Storage

Smaller form factor (e.g., 2.5 inch) hard disk drives often consume less power per gigabyte than physically larger drives.[45] [46] Unlike hard disk drives, solid-state drives store data in flash memory or DRAM. With no moving parts, power consumption may be reduced somewhat for low-capacity flash-based devices.[47] [48]

In a recent case study, Fusion-io, manufacturer of solid state storage devices, managed to reduce the energy use and operating costs of MySpace data centers by 80% while increasing performance speeds beyond that which had been attainable via multiple hard disk drives in Raid 0.[49] [50] In response, MySpace was able to retire several of their servers.

As hard drive prices have fallen, storage farms have tended to increase in capacity to make more data available online. This includes archival and backup data that would formerly have been saved on tape or other offline storage. The increase in online storage has increased power consumption. Reducing the power consumed by large storage arrays, while still providing the benefits of online storage, is a subject of ongoing research.[51]

Video card

A fast GPU may be the largest power consumer in a computer.[52]

Energy-efficient display options include:

- No video card - use a shared terminal, shared thin client, or desktop sharing software if display required.
- Use motherboard video output - typically low 3D performance and low power.
- Select a GPU based on low idle power, average wattage, or performance per watt.

Display

CRT monitors typically use more power than LCD monitors. They also contain significant amounts of lead. LCD monitors typically use a cold-cathode fluorescent bulb to provide light for the display. Some newer displays use an array of light-emitting diodes (LEDs) in place of the fluorescent bulb, which reduces the amount of electricity used by the display.[53] Fluorescent back-lights also contain mercury, whereas LED back-lights do not.

Materials recycling

Recycling computing equipment can keep harmful materials such as lead, mercury, and hexavalent chromium out of landfills, and can also replace equipment that otherwise would need to be manufactured, saving further energy and emissions. Computer systems that have outlived their particular function can be re-purposed, or donated to various charities and non-profit organizations.[54] However, many charities have recently imposed minimum system requirements for donated equipment.[55] Additionally, parts from outdated systems may be salvaged and recycled through certain retail outlets[56] [57] and municipal or private recycling centers. Computing supplies, such as printer cartridges, paper, and batteries may be recycled as well.[58]

A drawback to many of these schemes is that computers gathered through recycling drives are often shipped to developing countries where environmental standards are less strict than in North America and Europe.[59] The Silicon Valley Toxics Coalition estimates that 80% of the post-consumer e-waste collected for recycling is shipped abroad to countries such as China and Pakistan.[60]

In 2011, the collection rate of e-waste is still very low, even in the most ecology-responsible countries like France. In this country, e-waste collection is still at a 14% annual rate between electronic equipments sold and e-waste collected for 2006 to 2009.[61]

The recycling of old computers raises an important privacy issue. The old storage devices still hold private information, such as emails, passwords, and credit card numbers, which can be recovered simply by someone's using software available freely on the Internet. Deletion of a file does not actually remove the file from the hard drive. Before recycling a computer, users should remove the hard drive, or hard drives if there is more than one, and physically destroy it or store it somewhere safe. There are some authorized hardware recycling companies to whom the computer may be given for recycling, and they typically sign a non-disclosure agreement.[62]

Telecommuting

Teleconferencing and telepresence technologies are often implemented in green computing initiatives. The advantages are many; increased worker satisfaction, reduction of greenhouse gas emissions related to travel, and increased profit margins as a result of lower overhead costs for office space, heat, lighting, etc. The savings are significant; the average annual energy consumption for U.S. office buildings is over 23 kilowatt hours per square foot, with heat, air conditioning and lighting accounting for 70% of all energy consumed.[63] Other related initiatives, such as hotelling, reduce the square footage per employee as workers reserve space only when they need it.[64] Many types of jobs, such as sales, consulting, and field service, integrate well with this technique.

Voice over IP (VoIP) reduces the telephony wiring infrastructure by sharing the existing Ethernet copper. VoIP and phone extension mobility also made hot desking more practical.

Education and certification

Green computing programs

Degree and postgraduate programs that provide training in a range of information technology concentrations along with sustainable strategies in an effort to educate students how to build and maintain systems while reducing its negative impact on the environment. The Australian National University (ANU) offers "ICT Sustainability" as part of its information technology and engineering masters programs[65] . Athabasca University offer a similar course "Green ICT Strategies"[66] , adapted from the ANU course notes[67] . In the UK, Leeds Metropolitan University offers an MSc Green Computing program in both full and part time access modes [68] .

Green computing certifications

Some certifications demonstrate that an individual has specific green computing knowledge, including:

- Green Computing Initiative - GCI offers the Certified Green Computing User Specialist (CGCUS), Certified Green Computing Architect (CGCA) and Certified Green Computing Professional (CGCP) certifications.[69]
- CompTIA Strata Green IT is designed for IT managers to show that they have good knowledge of green IT practices and methods and why it is important to incorporate them into an organization.
- Information Systems Examination Board (ISEB) Foundation Certificate in Green IT is appropriate for showing an overall understanding and awareness of green computing and where its implementation can be beneficial.
- Singapore Infocomm Technology Federation (SiTF) Singapore Certified Green IT Professional is an industry endorsed professional level certification offered with SiTF authorized training partners. Certification requires completion of a four day instructor-led core course, plus a one day elective from an authorized vendor.[70]
- Australian Computer Society (ACS) The ACS offers a certificate for "Green Technology Strategies" as part of the Computer Professional Education Program (CPEP). Award of a certificate requires completion of a 12 week e-learning course, with written assignments.[71]

See also

- Camara (charity) (Ireland)
- *Challenging the Chip*, a book about labor rights and environmental justice in the global electronics industry
- Desktop virtualization
- Data migration
- Digger gold
- e-cycling
- eDay, an electronic waste collection day in New Zealand
- Electronic Waste Recycling Act
- Energy Efficient Ethernet
- Energy consumption of computers in the USA
- Interconnect bottleneck
- IT energy management
- Minimalism (computing)
- Optical communication
- Optical fiber cable
- Optical interconnect
- Parallel optical interfaceICT Sustainability
- Plug computer
- Power factor
- Power usage effectiveness (PUE)
- Rebound effect (paradoxical negative effect)
- Restriction of Hazardous Substances Directive (RoHS)
- Standby power
- Sustainable Electronics Initiative (SEI)
- Time-sharing
- Thunderbolt (interface)
- Trashware
- Virtual application

References

[1] San Murugesan, "Harnessing Green IT: Principles and Practices," IEEE *IT Professional*, January–February 2008, pp 24-33.

[2] "TCO takes the initiative in comparative product testing" (http://www.boivie.se/index.php?page=2&lang=eng). 2008-05-03. . Retrieved 2008-05-03.

[3] Full report: OECD Working Party on the Information Economy. "Towards Green ICT strategies: Assessing Policies and Programmes on ICTs and the Environment" (http://www.oecd.org/dataoecd/47/12/42825130.pdf). . Summary: OECD Working Party on the Information Economy. "Executive summary of OECD report" (http://www.oecd.org/dataoecd/46/18/43044065.pdf). .

[4] Jones, Ernesta (2006-10-23). "EPA Announces New Computer Efficiency Requirements" (http://yosemite.epa.gov/opa/admpress.nsf/ a8f952395381d3968525701c005e65b5/113b0c0647fee41585257210006474f1!OpenDocument). U.S. EPA. . Retrieved 2007-09-18.

[5] Gardiner, Bryan (2007-02-22). "How Important Will New Energy Star Be for PC Makers?" (http://www.pcmag.com/article2/ 0,1759,2097558,00.asp). PC Magazine. . Retrieved 2007-09-18.

[6] "DIRECTIVE 2002/96/EC OF THE EUROPEAN PARLIAMENT AND OF THE COUNCIL" (http://eur-lex.europa.eu/LexUriServ/ LexUriServ.do?uri=OJ:L:2003:037:0024:0038:EN:PDF). Official Journal of the European Union. 2003-01-27. . Retrieved 2009-10-21.

[7] "State Legislation on E-Waste" (http://www.e-takeback.org/docs open/Toolkit_Legislators/state legislation/state_leg_main.htm). Electronics Take Back Coalition. 2008-03-20. . Retrieved 2008-03-08.

[8] "Secretary Chu Announces $47 Million to Improve Efficiency in Information Technology and Communications Sectors" (http://www. energy.gov/news2009/8491.htm) (Press release). U.S. Department of Energy. 2010-01-06. . Retrieved 2010-10-30.

[9] "Intel and Google Join with Dell, EDS, EPA, HP, IBM, Lenovo, Microsoft, PG&E, World Wildlife Fund and Others to Launch Climate Savers Computing Initiative" (http://www.climatesaverscomputing.org/program/press.html) (Press release). Business Wire. 2007-06-12. . Retrieved 2007-12-11.

[10] "What exactly is the Climate Savers Computing Initiative?" (http://www.climatesaverscomputing.org/program/index.html). Climate Savers Computing Initiative. 2007. . Retrieved 2007-12-11.

[11] "President Bush Requires Federal Agencies to Buy EPEAT Registered Green Electronic Products" (http://www.epeat.net/Docs/Bush Requires EPEAT (1-24-07).pdf) (PDF) (Press release). Green Electronics Council. 2007-01-24. . Retrieved 2007-09-20.

[12] "Executive Order: Strengthening Federal Environmental, Energy, and Transportation Management" (http://georgewbush-whitehouse. archives.gov/news/releases/2007/01/20070124-2.html) (Press release). The White House: Office of the Press Secretary. 2007-01-24. . Retrieved 2007-09-20.

[13] "Energy benchmarks: a detailed analysis (e-Energy 2006)" (http://portal.acm.org/citation.cfm?doid=1791314.1791336). ACM. ISBN 978-1-4503-0042-1. Meikel Poess, Raghunath Nambiar, Kushagra Vaid, John M. Stephens, Jr., Karl Huppler, Evan Haines. .

[14] Donnellan, Brian and Sheridan, Charles and Curry, Edward (Jan/Feb 2011). "A Capability Maturity Framework for Sustainable Information and Communication Technology" (http://ieeexplore.ieee.org/lpdocs/epic03/wrapper.htm?arnumber=5708282). *IEEE IT Professional* **13** (1): 33–40. .

[15] "The common sense of lean and green IT" (http://www.deloitte.co.uk/TMTPredictions/technology/ Green-and-lean-it-data-centre-efficiency.cfm). *Deloitte Technology Predictions*. .

[16] InfoWorld July 06, 2009; http://www.infoworld.com/d/green-it/ used-pc-strategy-passes-toxic-buck-300?_kip_ipx=1053322433-1267784052&_pxn=0

[17] GreenIT.fr Feb. 2011; http://www.greenit.fr/article/materiel/pc-de-bureau/quelle-est-l-empreinte-carbone-d-un-ordinateur-3478

[18] Simon Mingay, Gartner: 10 Key Elements of a 'Green IT' Strategy; http://www.onsitelasermedic.com/pdf/10_key_elements_greenIT. pdf.

[19] "Best Practices Guide for Energy-Efficient Data Center Design", prepared by the National Renewable Energy Laboratory for the U.S. Department of Energy, Federal Energy Management Program, March 2011. (http://www1.eere.energy.gov/femp/pdfs/ eedatacenterbestpractices.pdf)

[20] Koomey, Jonathon. "Growth in data center electricity use 2005 to 2010," Oakland, CA: Analytics Press. August 1. (http://www. analyticspress.com/datacenters.html)

[21] "Best Practices Guide for Energy-Efficient Data Center Design", prepared by the National Renewable Energy Laboratory for the U.S. Department of Energy, Federal Energy Management Program, March 2011. (http://www1.eere.energy.gov/femp/pdfs/ eedatacenterbestpractices.pdf)

[22] "Research reveals environmental impact of Google searches." (http://www.foxnews.com/story/0,2933,479127,00.html). *Fox News*. 2009-01-12. . Retrieved 2009-01-15.

[23] "Powering a Google search" (http://googleblog.blogspot.com/2009/01/powering-google-search.html). *Official Google Blog*. Google. . Retrieved 2009-10-01.

[24] http://www.greenit.fr

[25] "Office suite require 70 times more memory than 10 years ago." (http://www.greenit.fr/article/logiciels/ logiciel-la-cle-de-l-obsolescence-programmee-du-materiel-informatique-2748). *GreenIT.fr*. 2010-05-24. . Retrieved 2010-05-24.

[26] Reardon, Marguerite (August 18, 2009). "Energy-aware Internet routing coming soon" (http://news.cnet.com/ 8301-11128_3-10312408-54.html). . Retrieved August 19, 2009.

[27] Kurp, Patrick."Green Computing," *Communications of the ACM*51(10):11.

[28] "Windows 95 Power Management" (http://msdn.microsoft.com/en-us/library/ms810046.aspx). .

[29] "Windows power-saving options are, bizarrely, stored in HKEY_CURRENT_USER" (http://www.liv.ac.uk/csd/greenit/powerdown). .

[30] "How Windows XP Wasted $25 Billion of Energy" (http://www.treehugger.com/files/2006/11/how_windows_xp.php). 2006-11-21. . Retrieved 2005-11-21.

[31] "Windows Vista Power Management Changes" (http://download.microsoft.com/download/5/b/9/ 5b97017b-e28a-4bae-ba48-174cf47d23cd/CPA075_WH06.ppt). .

[32] "Windows 7 Processor Power Management" (http://www.microsoft.com/whdc/system/pnppwr/powermgmt/ProcPowerMgmtWin7. mspx). .

[33] "Windows 7 Timer Coalescing" (http://www.microsoft.com/whdc/system/pnppwr/powermgmt/TimerCoal.mspx). .

[34] "Power Management Software for Windows Workstations" (http://www.windowsitpro.com/article/buyers-guide/ Power-Management-Software-for-Windows-Workstations-Buyers-Guide.aspx). .

[35] "Energy Star Commercial Packages List" (http://www.energystar.gov/index.cfm?c=power_mgt.pr_power_mgt_comm_packages). .

[36] The Headmasters' and Headmistresses' Conference. "HMC: A Practical Guide to Sustainable Building for Schools" (http://www. hmcsustainability.org.uk/energy.html). .

[37] "PC Power Management Solutions" (http://itmanagersinbox.com/1399/pc-power-management-solutions). .

[38] "Why use software NightWatchman to turn your PCs off?" (http://features.techworld.com/green-it/3546/ why-use-software-nightwatchman-to-turn-your-pcs-off/). .

[39] "University of Oxford Low Carbon Project: Energy and the networked computing environment" (http://projects.oucs.ox.ac.uk/ lowcarbonict/conferences/conf-2.htm#providers). .

[40] "Forrester Study: Total Economic Impact of Faronics Power Save" (http://www.faronics.com/Faronics/Documents/ Study_PowerSave_Forrester_TEI_EN.pdf). .

[41] "1E upgrades NightWatchman, seeks to bring powermanagement to SMEs: Competitive landscape" (http://www.1e.com/Downloads/
        Articles/Published/451 - 1E - Market Development-020309.pdf). .

[42] Schuhmann, Daniel (2005-02-28). "Strong Showing: High-Performance Power Supply Units" (http://www.tomshardware.com/2005/02/
        28/strong_showing/page38.html). Tom's Hardware. . Retrieved 2007-09-18.

[43] 80 PLUS

[44] "Computer Key Product Criteria" (http://www.energystar.gov/index.cfm?c=computers.pr_crit_computers). Energy Star. 2007-07-20. .
        Retrieved 2007-09-17.

[45] Mike Chin (8 March 2004). "IS the Silent PC Future 2.5-inches wide?" (http://www.silentpcreview.com/article145-page1.html). .
        Retrieved 2008-08-02.

[46] Mike Chin (2002-09-18). "Recommended Hard Drives" (http://www.silentpcreview.com/article29-page2.html). . Retrieved 2008-08-02.

[47] Super Talent's 2.5" IDE Flash hard drive - The Tech Report - Page 13 (http://techreport.com/articles.x/10334/13)

[48] Power Consumption - Tom's Hardware: Conventional Hard Drive Obsoletism? Samsung's 32 GB Flash Drive Previewed (http://www.
        tomshardware.com/reviews/conventional-hard-drive-obsoletism,1324-5.html)

[49] http://community.fusionio.com/media/p/458/download.aspx

[50] (http://www.forbes.com/feeds/businesswire/2009/10/13/businesswire130131961.html)

[51] IBM chief engineer talks green storage (http://searchstorage.techtarget.com/news/article/0,289142,sid5_gci1255635,00.
        html?track=NL-52&ad=592882&asrc=EM_NLN_1637134&uid=6241617), SearchStorage - TechTarget

[52] http://www.xbitlabs.com/articles/video/display/power-noise.html X-bit labs: Faster, Quieter, Lower: Power Consumption and Noise
        Level of Contemporary Graphics Cards

[53] "Cree LED Backlight Solution Lowers Power Consumption of LCD Displays" (http://news.thomasnet.com/fullstory/464080).
        2005-05-23. . Retrieved 2007-09-17.

[54] Reuse your electronics through donation » Earth 911 (http://earth911.org/blog/2007/04/12/
        reuse-your-electronics-through-donation-to-schools-charities-and-nonprofit-organizations/)

[55] Delaney, John (2007-09-04). "15 Ways to Reinvent Your PC" (http://www.pcmag.com/article2/0,1895,2170255,00.asp). *PC Magazine*
        26 (17). .

[56] "Staples Launches Nationwide Computer and Office Technology Recycling Program" (http://investor.staples.com/phoenix.
        zhtml?c=96244&p=irol-newsArticle&ID=1004542&highlight=). Staples, Inc.. 2007-05-21. . Retrieved 2007-09-17.

[57] "Goodwill Teams with Electronic Recyclers to Recycle eWaste" (http://earth911.org/blog/2007/08/15/
        goodwill-teams-with-electronic-recyclers-to-recycle-ewaste/). Earth 911. 2007-08-15. . Retrieved 2007-09-17.

[58] Refilled ink cartridges, paper recycling, battery recycling

[59] Segan, Sascha (2007-10-02). "Green Tech: Reduce, Reuse, That's It" (http://www.pcmag.com/article2/0,2704,2183977,00.asp). *PC
        Magazine* **26** (19): 56. . Retrieved 2007-11-07.

[60] Royte, Elizabeth (2006). *Garbage Land: On the Secret Trail of Trash*. Back Bay Books. pp. 169–170. ISBN 0-316-73826-3.

[61] "e-waste take-back still very low" (http://www.greenit.fr/article/materiel/recyclage/
        deee-l-europe-revoit-ses-objectifs-a-la-hausse-3482). GreenIT.fr. 2011-02-09. . Retrieved 2011-02-09.

[62] MySecureCyberspace » Green Computing and Privacy Issues (http://www.mysecurecyberspace.com/articles/features/
        green-computing-and-privacy-issues.html)

[63] "EPA Office Building Energy Use Profile" (http://www.epa.gov/cleanenergy/documents/sector-meeting/4bi_officebuilding.pdf)
        (PDF). EPA. 2007-08-15. . Retrieved 2008-03-17.

[64] "What Is Green IT?" (http://energypriorities.com/entries/2007/06/what_is_green_it_data_centers.php). .

[65] "ICT Sustainability Course" (http://studyat.anu.edu.au/courses/COMP7310;details.html). Australian National University. 2012-01-01. .
        Retrieved 2012-02-05.

[66] "Green ICT Strategies (Revision 1)" (http://scis.athabascau.ca/graduate/comp635syllabus.php). Athabasca University. 2011-09-06. .
        Retrieved 2012-02-05.

[67] "ICT Sustainability: Assessment and Strategies for a Low Carbon Future" (http://www.tomw.net.au/ict_sustainability/). Tomw
        Communications Pty, Limited. 2011-09-06. . Retrieved 2012-02-05.

[68] "MSc Green Computing Course" (http://courses.leedsmet.ac.uk/greencomputing_msc). Leeds Metropolitan University. 2012-01-01. .
        Retrieved 2012-04-26.

[69] "Green Computing Initiative Certification" (http://www.sgauge.com/greenci/). Green Computing Initiative. . Retrieved 2012-02-05.

[70] "Singapore Certified Green IT Professional" (http://www.sitf.org.sg/sitfacademy/Programme_SCGP.html). Singapore Infocomm
        Technology Federation. 2012. . Retrieved 2012-02-05.

[71] "Green Technology Strategies Course" (http://www.acs.org.au/cpeprogram/index.cfm?action=show&conID=greenict). Australian
        Computer Society. 2011-11-29. . Retrieved 2012-02-05.

Institute_for_Business_Value

The IBM **Institute for Business Value** is a business research organization that focuses on managerial and economic issues faced by companies and governments around the world. It has offices in China, India, Ireland, Japan, The Netherlands, South Africa and The United States, and It publishes between 35 and 50 major studies each year.

History

The IBM Institute for Business Value grew out of earlier research programs at IBM that were part of the firm's management consulting organization. These predecessors included the IBM Institute for Knowledge Management (IKM) and the IBM e-business Innovation Institute (ebII). The IKM, founded in 1999, was a membership consortium of private and public organizations that conducted research on the managerial and organizational aspects of knowledge management. The ebII was formed in 2000 with a small group of IBM consultants doing research into the impact of current and future technologies on business issues across a wide range of industries. In August 2001, IBM acquired Mainspring, a small strategy consulting firm headquartered in Cambridge, Massachusetts, and in October combined its research capabilities with the IKM and ebII to create the IBM Institute for Business Value (IBV).

From its early years the IBV focused on topics relevant to senior executives. It began publishing studies on business trends and management issues in such core industries as banking, automotive, and telecommunications. Most of the studies were global in scope, although a few focused on issues specific to a region, such as Western Europe or North America. Additionally, the IBV began to conduct research concerning cross-industry issues, such as customer relationship management (CRM).

In 2002, IBM acquired PriceWaterhouseCoopers Consulting (PwCC), the consulting arm of PriceWaterhouseCoopers, thereby increasing IBM's consulting population by some 30,000 employees and augmenting its expertise in the business issues of over a dozen industries and key functions, such as financial management, and supply chain management. The IBV became part of the merged consulting organization, now known as IBM Global Business Services (GBS).

The inclusion of PwCC into IBM led to important changes in the IBV. First, the number of full-time consultants working at the IBV grew from about 35 in 2002 to about 55 by 2008 and with the addition of specialized groups in Japan, India and China to about 80 in 2011. Second, many came in with a combination of advanced academic research degrees and consulting experience. Third, the staff became more international, although two main offices were established: one in Cambridge, Massachusetts and the other in Amsterdam. Studies were now organized more expansively by industry and management function (e.g., finance, HR). One type of study brought over from PwCC was the "future agenda" that explored the economic and business trends affecting an industry over a 5 to 10 year period. Within several years, the IBV had published nearly a dozen of such studies. The staff also published books dealing with such issues as knowledge management, the electronics industry, and the role of IT in myriad American industries.

Key Activities

The IBV continues to conduct primary research, publish white papers, present research findings at industry conferences, and write bylined articles for trade and business publications. Global activities continue to expand through IBV satellite programs in China, Japan, India and South Africa. A high profile collaboration, begun in 2001, with the Economist Intelligence Unit (EIU) produces a ranking of scores of nations in their use of IT, called the E-Readiness Ranking. Published annually in the spring, by 2008 it was the longest-running review of economic, political, social and technological issues affecting countries around the world

The IBV also has continued and broadened a series of CxO studies originated by PwCC. The CxO studies conduct extensive (often involving over 1,000 participants from across the world and from a wide range of sectors) in-office surveys and interviews with senior executives around the globe – CEOs [1] [2] [3] [4] , Chief Information Officers (CIOs)[5] [6] , Chief Financial Officers (CFOs)[7] [8] [9] , Chief Human Resource Officers (CHROs)[10] [11] [12] , Chief Supply Chain Officers (CSCOs)[13] and Chief Marketing Officers (CMOs), – publishing the results every two years for CEOs and CIOs in May and every four years for CFOs, CHROs, CSCOs and CMOs in October.

External links

- Institute for Business Value [14]

References

[1] "Global CEO Study 2004: Your Turn" (http://www-935.ibm.com/services/in/index.wss/ibvstudy/bcs/x1021921?cntxt=x1021855). ibm.com. . Retrieved 26 July 2011.

[2] "Global CEO Study 2006: Expanding the Innovation Horizon" (http://www-935.ibm.com/services/au/bcs/html/bcs_ceostudy2006. html). ibm.com. . Retrieved 26 July 2011.

[3] "Global CEO Study 2008: The Enterprise of the Future" (http://www.ibm.com/ibm/ideasfromibm/us/ceo/20080505/). ibm.com. . Retrieved 26 July 2011.

[4] "Global CEO Study 2010: Capitalizing on Complexity" (http://www-935.ibm.com/services/us/ceo/ceostudy2010/index.html). ibm.com. . Retrieved 26 July 2011.

[5] "Global CIO Study 2009: The New Voice of the CIO" (http://www-304.ibm.com/businesscenter/cpe/download0/183490/ MM_CIO_Study.pdf). ibm.com. . Retrieved 26 July 2011.

[6] "Global CIO Study 2011: The Essential CIO" (http://www-935.ibm.com/services/c-suite/cio/study.html). ibm.com. . Retrieved 26 July 2011.

[7] "Global CFO Study 2005: The Agile CFO" (https://www-31.ibm.com/cn/services/bcs/iibv/pdf/finance_cfo1.pdf). ibm.com. . Retrieved 26 July 2011.

[8] "Global CFO Study 2007: Balancing Risk and Performance with a Balanced Finance Organization" (http://www-03.ibm.com/press/us/en/ pressrelease/22499.wss). ibm.com. . Retrieved 26 July 2011.

[9] "Global CFO Study 2010: The New Value Integrator" (http://www-935.ibm.com/services/us/gbs/bus/html/gbs-2010cfostudy.html). ibm.com. . Retrieved 26 July 2011.

[10] "Global CHRO Study 2005: The Capability Within" (http://www-31.ibm.com/cn/services/bcs/iibv/pdf/final_en.pdf). ibm.com. . Retrieved 26 July 2011.

[11] "Global CHRO Study 2008: Unlocking the DNA of the Adaptable Workforce" (http://www-935.ibm.com/services/us/gbs/bus/html/ 2008ghcs.html). ibm.com. . Retrieved 26 July 2011.

[12] "Global CHRO Study 2010: Working beyond Borders" (http://www-935.ibm.com/services/c-suite/chro/study.html). ibm.com. . Retrieved 26 July 2011.

[13] "Global CSCO Study 2010: The Smarter Supply Chain of the Future" (http://www-935.ibm.com/services/us/gbs/bus/html/ gbs-csco-study.html). ibm.com. . Retrieved 26 July 2011.

[14] http://www-935.ibm.com/services/us/gbs/thoughtleadership/

Article Sources and Contributors

E-readiness *Source*: http://en.wikipedia.org/w/index.php?title=E-readiness *Contributors*: Abdel Hameed Nawar, Andrwsc, Andy Marchbanks, Cambrasa, Cherlin, Cybercobra, Desmond Tsui, Elockid, Equilibrial, GTBacchus, HenryLi, Hermógenes Teixeira Pinto Filho, Heroeswithmetaphors, Hinalover, Ictlogist, Jhattara, Jjhcap99, Johntex, Joseph Solis in Australia, Jpbowen, Kbh3rd, Ligulem, Neo-Jay, RHaworth, Redking7, Sjjupadhyay, Tikiwont, 28 anonymous edits

Economist_Intelligence_Unit *Source*: http://en.wikipedia.org/w/index.php?title=Economist_Intelligence_Unit *Contributors*: Adoniscik, Alan.ca, Andy Marchbanks, Andycjp, Bevo74, Brandon.colbert, Brianmulligan, Coat of Arms, Cocoaguy, Ctconnolly, Cybercobra, Darthbreather, Fadesga, FayssalF, HamburgerRadio, Indon, Innerproduct, Jonathan.s.kt, Kaji132007, Kane5187, LilHelpa, MER-C, Margana, Miguel in Portugal, MorrisGregorian, Neutrality, Pisshead0909, RJHall, Retired username, Shell Kinney, Shortride, Si sanett, Strawinthewind, UnitedStatesian, Vadimka, Yvwv, 41 anonymous edits

Information_and_communications_technology *Source*: http://en.wikipedia.org/w/index.php?title=Information_and_communications_technology *Contributors*: -iNu-, 28bytes, Aengland2, Ahunt, Alansohn, Allens, Andy4789, Anniefrankly, Apparition11, AseemGuptaDxb, Baldie, BioPupil, Bonjour68, BristolMan1979, CLW, Canto00, Cchase0405, Ciccioformaggio, Cirt, Cokoli, Connormah, Cvbrandoe, Deepak D'Souza, Delband, Deworrall, Discospinster, Dochar, Dommhegg, Drmies, Dumishka, EggNogInTheMorning, Eskadenia 2000, Espoo, Fæ, GEBStgo, Gabriel Kielland, Gaius Cornelius, Globaltka, Gonzjo, Hallenrm, Headroute, Hoo man, InfoCmplx, Informationist1, Isarra, Isheden, JDP90, JaGa, JanetteDoe, Jarble, Jean.julius, Jeremy Butler, Jim1138, Jmccormac, Joewski, Johnuniq, Jojhutton, JonH, Joshua Gyamfi, Jérôme, Kasjanek21, Kattmamma, Kevyn, Khvalamde, Knverma, Kocio, Kozuch, Kthapelo, Kuru, LachlanA, Lawelmor, Leszek Jańczuk, Leuko, Liampa, Lkopeter, Mahanga, Makru, Malcolmrlee, Mandarax, Marc Kupper, Mark91, Materialscientist, Mattyn 210, Mcarroll4, Mcrk, Meaghan, Meneer B, Mightymights, Mismilo79, Mjroots, MrOllie, Nasa-verve, Nhantdn, NickCT, Niranjandakua, Noformation, Ohnoitsjamie, PL290, Perohanych, Piano non troppo, Pluma, Prari, Preeti kumari, Pseudomonas, Pypo, Qwfp, RA0808, RHaworth, Ragimiri, Rangoon11, Raspalchima, Raywil, Reach Out to the Truth, Redirect fixer, Rich Farmbrough, Robangel, Roundtheworld, Rulerofutumno, Saintsmile, Santoshbhatt.nepal, Sapphire82, Semeiotike, Shaysom09, Siddhant, Skarebo, Starlemusique, Suffosion of Yellow, Ѕәэв,uıʞoɯ, The Thing That Should Not Be, Thecheesykid, Tide rolls, Tomchiukc, Trusilver, Umni2, V3rbum, Vashtihorvat, Warp9wb, Wavelength, WikHead, Wikipelli, Wine Guy, Wynne0, Yeng-Wang-Yeh, Yk Yk Yk, Ynhockey, Yupik, Yworo, Zhou Yu, Ztrawhcs, Zzuuzz, 261 anonymous edits

United_Nations_Public_Administration_Network *Source*: http://en.wikipedia.org/w/index.php?title=United_Nations_Public_Administration_Network *Contributors*: Abdel Hameed Nawar, Courcelles, Emeraude, Good Olfactory, Kku, Koavf, MasterOfHisOwnDomain, Redeagle688, Rory096, TigerShark, Unpan, Visviva, Zappa711, 22 anonymous edits

World_Bank *Source*: http://en.wikipedia.org/w/index.php?title=World_Bank *Contributors*: -Majestic-, 213.253.39.xxx, 3mta3, A2682159, AJR, Aakash58, Abpayne, Acerperi, Acsenray, AdamRaizen, Agentile27, Alan Liefting, Alansohn, Aleksd, Alexander Matthew Ball, AlexiusHoratius, Alvis, Amarteifio, Ameliorate!, Anaxial, Andrzej Kmicic, Andycjp, Anirvan, Ark25, Ask123, Auntof6, AutoGyro, Avadlehunter456, Barek, Beland, Bento00, Bigturtle, Bjorn Martiz, Blanchardb, Bobo192, Bobrayner, Bongwarrior, Bonzolive, Boulaur, Burrburr, Byelf2007, CCCarson, CZmarlin, Cab88, Calabe1992, CalendarWatcher, Capricorn42, CaroleHenson, Catgut, Ched Davis, Chinneeb, Chokoboii, Chowbok, ChrisLamb, Christoph.sta, Cntras, Cometstyles, Conversion script, Cookooji, Cool anurag upadhyay, Cornellrockey, Courcelles, Cretog8, Cromwellt, Cwescott, Cynwolfe, Dakshinamurthy, Dalstadt, Daniel Quinlan, Dann, DarkAudit, Davecrosby uk, David Parker, DavidHoag, DavidLevinson, Dawn Bard, Dchall1, Debresser, Decent 11, Delfindakila, Delldot, Delldot on a public computer, DerKaiser, Development now, DiceDiceBaby, Diogenes zosimus, Diya9005, Diza, DocWatson42, Docu, Douwe20, Drgregmartin, Drmies, Dswislow, Duffman, EWikist, Earthlyreason, Edmilne, Educatedseacucumber, Edward, Eik Corell, Elen of the Roads, Emhoo, Enauspeaker, Enigmaman, Epbr123, Erianna, Ericl234, Erik9, EuroTom, Eurodad, Evil Monkey, F345, Fanyavizuri, Fayenatic london, Fdp10, Fenhaus, FeydHuxtable, Fieldday-sunday, Flubeca, Former user, Foxhound66, FrankTobia, Franz weber, Fredrik, Futuro13, Gabbe, Gadfium, Galapagos999, Gary King, Gioto, Globalwikireader, Gobonobo, Godvarryu, Gogo Dodo, Good Olfactory, Gorba, Graham87, Grant76, Gritzko, Ground Zero, Gtstricky, Guaka, Guppy, Gurch, Guusbosman, Hard Sin, Hemanshu, Henigushi, Heron, Hibernian, II MusLiM HyBRiD II, Ichwan Palongengi, Igoldste, Ih1991, Iloveandrea, Imagine&Engage, ImperfectlyInformed, Inliten, Itai, J.delanoy, JForget, JJallman, Jaakkoh, Jacob Lundberg, Jakatrpe, Jambalay, JamesAM, JamesBWatson, Jan1nad, January, Japs 88, JenLouise, Jiang, JinJian, Jmill1990, JodyB, Joe 042293, Joey9630, John, John Quiggin, John Vandenberg, JonHarder, Joseph Solis in Australia, Jovianeye, Jrtayloriv, JudeFinisterra, Julius Sahara, Just plain Bill, Justin.sharber, KFP, Kaaveh Ahangar, Kaboch, Karthickbabub, Katalaveno, Kavafy, Keilana, Keith-264, KiViki, Kingpomba, KnowledgeOfSelf, Koavf, Kokiri, Koyaanis Qatsi, Kozuch, Krunchky, Kuool12, Kwertii, KyuuA4, LOL, Lapsed Pacifist, Lemur12, Leonard^Bloom, Levineps, Lihaas, LindsayH, Little Samourai, Lokpest, Lordevelynderothschild, Loren.wilton, Lotje, Louddrummersx2, Ludo716, Lygophile, MCMLXXXVII, MDCore, MaGioZal, MaNeMeBasat, Mac, Maikoherajin, Malcolm Farmer, Man It's So Loud In Here, Marc-Olivier Pagé, Marek69, Marlinz144, Marta2009, Martyn bg, Masalih, Materialscientist, Matttorpto, Maunus, Mav, Maziotis, McAlcibiades, Mcauburn, Mendali, Mgllama, Miguelaaron, Mike Rosoft, Mintleaf, Misterx2000, Mitsuhirato, Montirakrich, Mophoplz, MrNiceGuy1113, Mschlindwein, Muck mcuk, Naddy, Namayan, Nandt1, Naniwako, Ndenison, Nerd, Neutral Milk Hotel, Nick Number, Nick UA, Nifky?, Nikkimaria, Nirvana888, Nobs01, Nopetro, O-Jay, Obamizacion, Ohconfucius, Ojay123, Olivier, Orange Suede Sofa, Ospalh, Otherwho, Ottawahitech, OverlordQ, PFHLai, Palix, Panarchy, Passionless, Patos, PatrickFlaherty, Patxi lurra, Paul789a, Pbelli, PenguiN42, Pexise, Phd8511, Pichulotote2, Pilarloren, Plutonics, Porius1, Postdlf, Prevalis, PrimeHunter, Pushpin, Rn'B, Ragnarthebloodaxe, Rangasyd, Rank-one map, Rd232, Rd232 public, Reconsider the static, Redfarmer, Registreernu, Rfc1394, Rich Farmbrough, Rich257, RickRowden, Rillian, Rishabhpodar, Rmccue, RobOWU, Robinh, Ruiz, Rédacteur Tibet, SD6-Agent, SJK, Saebvn, Sapovadia, Saturnight, Satyanandapatel, ScapegoatVandal, Scartboy, SchnitzelMannGreek, Scientus, Seaphoto, Senator Palpatine, Shadowjams, ShajiA, Shawn789, Shayansamii, Shimwell, Shrigley, Sibi antony, Signorini.j, Simon Shek, SimonP, Slawojarek, Slrubenstein, Snoyes, Soccerdude00092, Soerfm, Sol Blue, Somesparetime, Sorrowsolve, SpeedyGonsales, SpikeToronto, Spiraldays, Staeiou, Stormwriter, Subversive.sound, Suiosadef, Susan Chan, Susan Mason, SusanLesch, SuzanneKn, TakuyaMurata, Temperingking, The Baroness of Morden, The Thing That Should Not Be, Therequiembellishere, Thingg, Thomas Yeardly, Throwaway85, Tiddly Tom, Tide rolls, Tigga en, Tiggerjay, Timwi, Tktru, Tomlillis, Torren53, Transposeinverse, Triona, Tristanb, Trypa, Tyrol5, Ui9sd, Ultraexactzz, UnicornTapestry, UpstateNYer, Urhixidur, Utcursch, Van helsing, Versus22, Vinsci, Voidxor, Vuvar1, WEALLDROP, Wavelength, Wayne Slam, Whiskey in the Jar, Wikibubbles, Wikiuser100, Willtron, Winchelsea, Wizzard, Wk muriithi, Yerpo, Ykhwong, Youssefsan, Yvwv, ZBull84, ZabMilenko, Zach Vega, Zain Ebrahim111, Zarcadia, Zuracech lordum, Zweifel, Zzuuzz, Александър, احمدغلامي.24, 714 anonymous edits

Digital_divide *Source*: http://en.wikipedia.org/w/index.php?title=Digital_divide *Contributors*: 01swinj, 1exec1, 5Celcious, 67-21-48-122, A5y, AZ5461, Aacarling, Aaron Brenneman, Abrech, Acarvin, Aches22, Achitnis, Achresto, Adamskio, Adashiel, Ahoerstemeier, Aitias, Aka042, Alan Liefting, Alansohn, Alekperova, AlexP, Alimyster9, Altenmann, Anand Karia, Andrew D White, Andy M. Wang, Andy120290, Andycjp, Animum, Anna Lincoln, Anngene, Appraiser, Ar54454, ArCePi, Aronzak, Arthur Rubin, Astonj s, Athens22, Auntof6, Avalon, Babym1424, Balthazarduju, Baristarim, Bartonnator, BatchProcessor, Baxrob, Beland, Benjamin Sherwin, Bg71792, Bhadani, Big Brother 1984, Blandings, Blendus, Bluedelta, Bob bobato, Bobo192, Bomac, Boo2989, Boris Barowski, Braders17, Brittbyers, Calliopejen1, Caltas, Camilo Sanchez, Cantus, Cbissell, CharlotteWebb, Chaserokkit, Chris the speller, Chrislems, Chrissi, Christophermatar, Ckatz, Clodagh.veale, Closedmouth, CloudNine, Comm12-david, CommGuy75, CommonsDelinker, Computerjoe, Cooldudeomg, Corky842, Courcelles, Cowboy pup, Creationist Phil, Crscrs, Cultofpj, DGtal, DMG413, Dan100, DancingPenguin, Dangerousnerd, Darth Panda, Dawnseeker2000, DebbieV, DerHexer, Dialectric, DigitalDev, Diligent Terrier, Discospinster, Dlohcierekim, Doc glasgow, Dogposter, Dominique212, Donatrip, Donreed, DorianS1, Dpr, Dr Gangrene, Draco Hunter, Drmies, Drmoores, Dueyfinster, Duk, Dusik, Dycedarg, E0steven, ESKog, Edward, EightiesRocker, El C, Elonka, Entropy, Epbr123, Erianna, EstebanF, Euractiv, Everyking, Excirial, Falcon8765, Fayenatic london, Ffaker, Fieldday-sunday, Flamingspinach, Floorwalker, Foxmulder, FreplySpang, Frscght, Fvw, GEBStgo, Gabriele.wiki, Gaius Cornelius, Galanv, Gary King, Gdstokes, Geastes, Gikü, Gimboid13, Gronky, Grosscha, Ground Zero, Haakon, HalJor, Hanrahan, Hemmingsen, Henryhjkim, Hephaestos, HexaChord, Hkfiedler, Hmains, ICTlogist, INVERTED, IatrosH, Idkanything, Ikonact, Imoeng, Infin8dawg, InfoCmplx, Ireneswu, Iroony, Istudentstgeorge, Ixfd64, J.delanoy, JLipshin, JREL, Jaded.snowflake, Jaideraf, JamieS93, JanetteDoe, Jarebear415, Jarinya gate, Jasper Deng, Jazzwick, Jbmurray, JiFish, Jim.henderson, Jizobizo, Jncraton, JoanneB, Jobber, Joe393987, Johnleemk, Johnteslade, Johntex, JonTheMon, Joolz, Jose Icaza, Joseph Solis in Australia, Jpbowen, Jugger90, Juliancolton, Jusdafax, Justin401, Jwild, Jwmsfobcn, KVDP, Kairosurfing, Kaldari, Kalogeropoulos, Katana0182, Katrinal24, Kattmamma, Kb1282, Kbrose, Kcordina, Keibr, Keith Lehwald, Keithlard, Kerim Friedman, Khargas, Klausjogi, Knowledgemania, Kocio, Kokomo5, Komitsuki, Kozuch, Kraigag, Kreachure, Kukini, La Pianista, Landon496, LeoRubin64, Lesmanalim, Leszek Jańczuk, Letstalk, Levineps, Lexus2514, Leyo, LightAnkh, Lightdarkness, Lights, Linkspamremover, Lmatt, Logan, Lotje, Lquilter, Lucasjohnson, Lucidwave, Luk, Lupo, MC MasterChef, MER-C, MRLC94, MStraw, MToolen, Mac, Malhonen, Manishearth, MarQ, MarekTT, MarkS, Markdask, Mcharle28, Mcnuttjg, Mechanical digger, Mediastu, Mentifisto, Meredithwhite, Meredyth, Michaelbusch, Mike Rosoft, Mikecb, Mohit, Moneyerwin3, Monica Linville, Moriori, Morton devonshire, MrOllie, MrTsunami, MsMaam, Mukadderat, Musical Linguist, Mv60046, NCurse, Natalie Erin, Nauticashades, NawlinWiki, Niceguyedc, Nick, Nicoleb19, Nifky?, Nigosh, Nikai, Niteowlneils, Nleamy, Noommos, Notcaptainhook1, Npercz, Ocaasi, Oldiowl, Olivierd, Omicronpersei8, Orange Suede Sofa, OrbitOne, Otterfan, PAHGOO, Padmin22, Pandasandpenguins, Pasquale, Patrickdavidson, Pb30, Pearle, Pedro Aguiar, Perohanych, Peruvianllama, Petermshane, Pgan002, Piano non troppo, Piotrus, Planterns, Pne, Polyamorph, Pprabakar, Priceman86, Pyrospirit, Qsutton, Quess, Qwyrxian, R'n'B, Rachellechong, RainbowOfLight, Rbarreira, Recognizance, RedWolf, Reg porter, Rell Canis, RexNL, Riana, Rich Farmbrough, Rjakew, Rjwilmsi, Robth, RomanceNovel7, Ronz, RoyBoy, Roystonea, Rpsimbol, Rumer06, Russeasby, S401, SCU2013, SCcomm12, Sadalmelik, Salsa Shark, Sammmiiiiiii20000, Sampo, SasiSasi, Scepia, Sceptre, Schzmo, Schoolssout, Seaphoto, Searchme, Sg nash, Shadowjams, Shaky3001, Shell Kinney, Shi Hou, Shii, Shinmawa, Silasdog, Simesa, SiobhanHansa, Sjakkalle, Sjjupadhyay, SlackerMom, Slady, Sleuth21, Smileyface11945, Smoktambam, Solimansoliman, SophiaAlex92, SpaceFlight89, Spacesoon, Spilbrick, SpuriousQ, Sross (Public Policy), Srpnor, St@teaction, Stephan Leeds, Stephluke, Steven Walling, Stevendyson90, Stevenfruitsmaak, Stevietheman, Student1028, Student37829, Stustu12, Sysiphe, THEN WHO WAS PHONE?, Tainter, Taranet, Tbhotch, Technobadger, Technofuture, Tedickey, The Rambling Man, The Thing That Should Not Be, The wub, TheAMmollusc, TheoClarke, Thingg, Think outside the box, Thorseth, Tomos, Topbanana, Trainthh, Traveler100, Treyt021, Tripower, Unyoyega, Veinor, Velella, Vespristiano, Viridian, Vishuforall, Vrenator, W Nowicki, Waggers, Wavelength, Wayward, Weregerbil, Weser, Whitneyanne, WikHead, Wiki alf, WikiDan61, WikiDao, Wikiklrsc, Wikimickey49, Wlmg, Wmt, Wordmonkey, WordyGirl90, Wynnj26, Yakushima, Yamachewyow, Youssefsan, ZimZalaBim, Zzuuzz, , 1104 anonymous edits

Green_computing *Source*: http://en.wikipedia.org/w/index.php?title=Green_computing *Contributors*: 28bytes, Adolphus79, Aiyizo, Akira Tomosuke, Alan Liefting, Alansohn, Aleph Infinity, Aliavni, Alpha 4615, Ameliorate!, Analyst10, Anand akela, Andreig84, ArnoldReinhold, Aspevents, BWirth, Behun, Beland, BenignEditor, Bill.albing, BloomingtonSky, Bruce404, C777, CFilm, Cdamcke, Church of emacs, Ckatz, CoupDeMistral, Cut1664, Cwolfsheep, DARTH SIDIOUS 2, Danderson68, David44357, Davidobrisbane, Dmalis, Dman727, Doctorfluffy, Dpanda,

Image Sources, Licenses and Contributors

Image:EIU e-readiness 2008.svg *Source*: http://en.wikipedia.org/w/index.php?title=File:EIU_e-readiness_2008.svg *License*: unknown *Contributors*: User:Jhattara

File:Flag of Denmark.svg *Source*: http://en.wikipedia.org/w/index.php?title=File:Flag_of_Denmark.svg *License*: unknown *Contributors*: User:Madden

File:Flag of Sweden.svg *Source*: http://en.wikipedia.org/w/index.php?title=File:Flag_of_Sweden.svg *License*: unknown *Contributors*: Anomie

File:Flag of the Netherlands.svg *Source*: http://en.wikipedia.org/w/index.php?title=File:Flag_of_the_Netherlands.svg *License*: unknown *Contributors*: User:Zscout370

File:Flag of Norway.svg *Source*: http://en.wikipedia.org/w/index.php?title=File:Flag_of_Norway.svg *License*: unknown *Contributors*: User:Dbenbenn

File:Flag of the United States.svg *Source*: http://en.wikipedia.org/w/index.php?title=File:Flag_of_the_United_States.svg *License*: unknown *Contributors*: Anomie

File:Flag of Australia.svg *Source*: http://en.wikipedia.org/w/index.php?title=File:Flag_of_Australia.svg *License*: unknown *Contributors*: Anomie, Mifter

File:Flag of Singapore.svg *Source*: http://en.wikipedia.org/w/index.php?title=File:Flag_of_Singapore.svg *License*: unknown *Contributors*: Various

File:Flag of Hong Kong.svg *Source*: http://en.wikipedia.org/w/index.php?title=File:Flag_of_Hong_Kong.svg *License*: unknown *Contributors*: Designed by

File:Flag of Canada.svg *Source*: http://en.wikipedia.org/w/index.php?title=File:Flag_of_Canada.svg *License*: unknown *Contributors*: Anomie

File:Flag of Finland.svg *Source*: http://en.wikipedia.org/w/index.php?title=File:Flag_of_Finland.svg *License*: unknown *Contributors*: User:SKopp

File:Flag of New Zealand.svg *Source*: http://en.wikipedia.org/w/index.php?title=File:Flag_of_New_Zealand.svg *License*: unknown *Contributors*: Adabow, Adambro, Arria Belli, Avenue, Bawolff, Bjankuloski06en, ButterStick, Denelson83, Donk, Duduziq, EugeneZelenko, Fred J, Fry1989, Hugh Jass, Ibagli, Jusjih, Klemen Kocjancic, Mamndassan, Mattes, Nightstallion, O, Peeperman, Poromiami, Reisio, Rfc1394, Shizhao, Tabasco, Transparent Blue, Väsk, Xufanc, Zscout370, 35 anonymous edits

File:Flag of Switzerland.svg *Source*: http://en.wikipedia.org/w/index.php?title=File:Flag_of_Switzerland.svg *License*: unknown *Contributors*: User:-xfi-, User:Marc Mongenet, User:Zscout370

File:Flag of the United Kingdom.svg *Source*: http://en.wikipedia.org/w/index.php?title=File:Flag_of_the_United_Kingdom.svg *License*: unknown *Contributors*: Anomie, Mifter

File:Flag of Austria.svg *Source*: http://en.wikipedia.org/w/index.php?title=File:Flag_of_Austria.svg *License*: unknown *Contributors*: User:SKopp

File:Flag of France.svg *Source*: http://en.wikipedia.org/w/index.php?title=File:Flag_of_France.svg *License*: unknown *Contributors*: Anomie

File:Flag of the Republic of China.svg *Source*: http://en.wikipedia.org/w/index.php?title=File:Flag_of_the_Republic_of_China.svg *License*: unknown *Contributors*: 555, Abner1069, Bestalex, Bigmorr, Denelson83, Ed veg, Gzdavidwong, Herbythyme, Isletakee, Kakoui, Kallerna, Kibinsky, Mattes, Mizunoryu, Neq00, Nickpo, Nightstallion, Odder, Pymouss, R.O.C, Reisio, Reuvenk, Rkt2312, Rocket000, Runningfridgesrule, Samwingkit, Sasha Krotov, Shizhao, Tabasco, Vzb83, Wrightbus, ZooFari, Zscout370, 74 anonymous edits

File:Flag of Germany.svg *Source*: http://en.wikipedia.org/w/index.php?title=File:Flag_of_Germany.svg *License*: unknown *Contributors*: Anomie

File:Flag of Ireland.svg *Source*: http://en.wikipedia.org/w/index.php?title=File:Flag_of_Ireland.svg *License*: unknown *Contributors*: User:SKopp

File:Flag of South Korea.svg *Source*: http://en.wikipedia.org/w/index.php?title=File:Flag_of_South_Korea.svg *License*: unknown *Contributors*: Various

File:Flag of Belgium (civil).svg *Source*: http://en.wikipedia.org/w/index.php?title=File:Flag_of_Belgium_(civil).svg *License*: unknown *Contributors*: Bean49, David Descamps, Dbenbenn, Denelson83, Evanc0912, Fry1989, Gabriel trzy, Howcome, IvanOS, Ms2ger, Nightstallion, Oreo Priest, Rocket000, Rodejong, Sir Iain, ThomasPusch, Warddr, Zscout370, 4 anonymous edits

File:Flag of Bermuda.svg *Source*: http://en.wikipedia.org/w/index.php?title=File:Flag_of_Bermuda.svg *License*: unknown *Contributors*: User:Cronholm144, user:Nameneko, user:Nightstallion

File:Flag of Japan.svg *Source*: http://en.wikipedia.org/w/index.php?title=File:Flag_of_Japan.svg *License*: unknown *Contributors*: Anomie

File:Flag of Malta.svg *Source*: http://en.wikipedia.org/w/index.php?title=File:Flag_of_Malta.svg *License*: unknown *Contributors*: Alkari, Fry1989, Gabbe, Homo lupus, Klemen Kocjancic, Liftarn, Mattes, Meno25, Nightstallion, Peeperman, Pumbaa80, Ratatosk, Rodejong, Zscout370, 5 anonymous edits

File:Flag of Estonia.svg *Source*: http://en.wikipedia.org/w/index.php?title=File:Flag_of_Estonia.svg *License*: unknown *Contributors*: User:PeepP, User:SKopp

File:Flag of Spain.svg *Source*: http://en.wikipedia.org/w/index.php?title=File:Flag_of_Spain.svg *License*: unknown *Contributors*: Anomie

File:Flag of Italy.svg *Source*: http://en.wikipedia.org/w/index.php?title=File:Flag_of_Italy.svg *License*: unknown *Contributors*: Anomie

File:Flag of Israel.svg *Source*: http://en.wikipedia.org/w/index.php?title=File:Flag_of_Israel.svg *License*: unknown *Contributors*: AnonMoos, Bastique, Bobika, Brown spite, Captain Zizi, Cerveaugenie, Drork, Etams, Fred J, Fry1989, Geagea, Himasaram, Homo lupus, Humus sapiens, Klemen Kocjancic, Komando12, Kookaburra, Luispihormiguero, Madden, Neq00, NielsF, Nightstallion, Oren neu dag, Patstuart, PeeJay2K3, Pumbaa80, Ramiy, Reisio, Rodejong, SKopp, Sceptic, SomeDudeWithAUserName, Technion, Typhix, Valentinian, Yellow up, Zscout370, 31 anonymous edits

File:Flag of Portugal.svg *Source*: http://en.wikipedia.org/w/index.php?title=File:Flag_of_Portugal.svg *License*: unknown *Contributors*: User:Nightstallion

File:Flag of Slovenia.svg *Source*: http://en.wikipedia.org/w/index.php?title=File:Flag_of_Slovenia.svg *License*: unknown *Contributors*: User:SKopp, User:Vzb83, User:Zscout370

File:Flag of Chile.svg *Source*: http://en.wikipedia.org/w/index.php?title=File:Flag_of_Chile.svg *License*: unknown *Contributors*: Alkari, B1mbo, David Newton, Dbenbenn, Denelson83, ElmA, Er Komandante, Fibonacci, Fry1989, Fsopolonezcaro, Herbythyme, Huhsunqu, Kallerna, Kanonkas, Klemen Kocjancic, Kyro, Mattes, Mozzan, Nagy, Nightstallion, Piastu, Pixeltoo, Pumbaa80, SKopp, Sarang, Srtxg, Sterling.M.Archer, Str4nd, Ultratomio, Vzb83, Xarucoponce, Yakoo, Yonatanh, Zscout370, 42 anonymous edits

File:Flag of the Czech Republic.svg *Source*: http://en.wikipedia.org/w/index.php?title=File:Flag_of_the_Czech_Republic.svg *License*: unknown *Contributors*: special commission (of code): SVG version by cs:-xfi-. Colors according to Appendix No. 3 of czech legal Act 3/1993. cs:Zirland.

File:Flag of Lithuania.svg *Source*: http://en.wikipedia.org/w/index.php?title=File:Flag_of_Lithuania.svg *License*: unknown *Contributors*: User:SKopp

File:Flag of Greece.svg *Source*: http://en.wikipedia.org/w/index.php?title=File:Flag_of_Greece.svg *License*: unknown *Contributors*: (of code) (talk)

File:Flag of the United Arab Emirates.svg *Source*: http://en.wikipedia.org/w/index.php?title=File:Flag_of_the_United_Arab_Emirates.svg *License*: unknown *Contributors*: Anime Addict AA, Avala, Dbenbenn, Duduziq, F l a n k e r, Fry1989, Fukaumi, Gryffindor, Guanaco, Homo lupus, Kacir, Klemen Kocjancic, Krun, Madden, Neq00, Nightstallion, Piccadilly Circus, Pmsyyz, RamzyAbueita, 4 anonymous edits

File:Flag of Hungary.svg *Source*: http://en.wikipedia.org/w/index.php?title=File:Flag_of_Hungary.svg *License*: unknown *Contributors*: User:SKopp

File:Flag of Slovakia.svg *Source*: http://en.wikipedia.org/w/index.php?title=File:Flag_of_Slovakia.svg *License*: unknown *Contributors*: User:SKopp

File:Flag of Latvia.svg *Source*: http://en.wikipedia.org/w/index.php?title=File:Flag_of_Latvia.svg *License*: unknown *Contributors*: User:SKopp

File:Flag of Malaysia.svg *Source*: http://en.wikipedia.org/w/index.php?title=File:Flag_of_Malaysia.svg *License*: unknown *Contributors*: User:SKopp

File:Flag of Poland.svg *Source*: http://en.wikipedia.org/w/index.php?title=File:Flag_of_Poland.svg *License*: unknown *Contributors*: Anomie, Mifter

File:Flag of Mexico.svg *Source*: http://en.wikipedia.org/w/index.php?title=File:Flag_of_Mexico.svg *License*: unknown *Contributors*: User:AlexCovarrubias

File:Flag of South Africa.svg *Source*: http://en.wikipedia.org/w/index.php?title=File:Flag_of_South_Africa.svg *License*: unknown *Contributors*: Adriaan, Anime Addict AA, AnonMoos, BRUTE, Daemonic Kangaroo, Dnik, Duduziq, Dzordzm, Fry1989, Homo lupus, Jappalang, Juliancolton, Kam Solusar, Klemen Kocjancic, Klymene, Lexxyy, Mahahahaneapneap, Manuelt15, Moviedefender, NeverDoING, Ninane, Poznaniak, Przemub, SKopp, ThePCKid, ThomasPusch, Tvdm, Ultratomio, Vzb83, Zscout370, 35 anonymous edits

File:Flag of Brazil.svg *Source*: http://en.wikipedia.org/w/index.php?title=File:Flag_of_Brazil.svg *License*: unknown *Contributors*: Anomie

File:Flag of Turkey.svg *Source*: http://en.wikipedia.org/w/index.php?title=File:Flag_of_Turkey.svg *License*: unknown *Contributors*: User:Dbenbenn

File:Flag of Jamaica.svg *Source*: http://en.wikipedia.org/w/index.php?title=File:Flag_of_Jamaica.svg *License*: unknown *Contributors*: Anime Addict AA, Boricuaeddie, Davepape, Duduziq, Fred J, Fry1989, Herbythyme, KBarnett, Kilom691, Klemen Kocjancic, Kounoupidi, Körnerbrötchen, Ludger1961, Mattes, Nishkid64, Odder, Reisio, SKopp, Sarang, The Evil IP address, Wknight94, 27 anonymous edits

File:Flag of Argentina.svg *Source*: http://en.wikipedia.org/w/index.php?title=File:Flag_of_Argentina.svg *License*: unknown *Contributors*: User:Dbenbenn

File:Flag of Trinidad and Tobago.svg *Source*: http://en.wikipedia.org/w/index.php?title=File:Flag_of_Trinidad_and_Tobago.svg *License*: unknown *Contributors*: AnonMoos, Boricuaeddie, Duduziq, Enbéká, Fry1989, Homo lupus, Klemen Kocjancic, Madden, Mattes, Nagy, Neq00, Nightstallion, Pumbaa80, SKopp, Tomia, 10 anonymous edits

File:Flag of Bulgaria.svg *Source*: http://en.wikipedia.org/w/index.php?title=File:Flag_of_Bulgaria.svg *License*: unknown *Contributors*: User:SKopp

File:Flag of Romania.svg *Source*: http://en.wikipedia.org/w/index.php?title=File:Flag_of_Romania.svg *License*: unknown *Contributors*: User:AdiJapan

File:Flag of Thailand.svg *Source*: http://en.wikipedia.org/w/index.php?title=File:Flag_of_Thailand.svg *License*: unknown *Contributors*: User:Zscout370

File:Flag of Jordan.svg *Source*: http://en.wikipedia.org/w/index.php?title=File:Flag_of_Jordan.svg *License*: unknown *Contributors*: User:SKopp

File:Flag of Saudi Arabia.svg *Source*: http://en.wikipedia.org/w/index.php?title=File:Flag_of_Saudi_Arabia.svg *License*: unknown *Contributors*: Unknown

File:Flag of Colombia.svg *Source*: http://en.wikipedia.org/w/index.php?title=File:Flag_of_Colombia.svg *License*: unknown *Contributors*: User:SKopp

File:Flag of Peru.svg *Source*: http://en.wikipedia.org/w/index.php?title=File:Flag_of_Peru.svg *License*: unknown *Contributors*: User:Dbenbenn

File:Flag of the Philippines.svg *Source*: http://en.wikipedia.org/w/index.php?title=File:Flag_of_the_Philippines.svg *License*: unknown *Contributors*: Aira Cutamora

File:Flag of Venezuela.svg *Source*: http://en.wikipedia.org/w/index.php?title=File:Flag_of_Venezuela.svg *License*: unknown *Contributors*: Alkari, Bastique, Denelson83, DerFussi, Fry1989, George McFinnigan, Herbythyme, Homo lupus, Huhsunqu, Infrogmation, K21edgo, Klemen Kocjancic, Ludger1961, Neq00, Nightstallion, Reisio, Rupert Pupkin, ThomasPusch, Vzb83, Wikisole, Zscout370, 12 anonymous edits

File:Flag of the People's Republic of China.svg *Source*: http://en.wikipedia.org/w/index.php?title=File:Flag_of_the_People's_Republic_of_China.svg *License*: unknown *Contributors*: User:Denelson83, User:SKopp, User:Shizhao, User:Zscout370

File:Flag of Egypt.svg *Source*: http://en.wikipedia.org/w/index.php?title=File:Flag_of_Egypt.svg *License*: unknown *Contributors*: Open Clip Art

File:Flag of India.svg *Source*: http://en.wikipedia.org/w/index.php?title=File:Flag_of_India.svg *License*: unknown *Contributors*: Anomie, Mifter

File:Flag of Russia.svg *Source*: http://en.wikipedia.org/w/index.php?title=File:Flag_of_Russia.svg *License*: unknown *Contributors*: Anomie

File:Flag of Ecuador.svg *Source*: http://en.wikipedia.org/w/index.php?title=File:Flag_of_Ecuador.svg *License*: unknown *Contributors*: President of the Republic of Ecuador, Zscout370

File:Flag of Nigeria.svg *Source*: http://en.wikipedia.org/w/index.php?title=File:Flag_of_Nigeria.svg *License*: unknown *Contributors*: User:Jhs

File:Flag of Ukraine.svg *Source*: http://en.wikipedia.org/w/index.php?title=File:Flag_of_Ukraine.svg *License*: unknown *Contributors*: User:Jon Harald Søby, User:Zscout370

File:Flag of Sri Lanka.svg *Source*: http://en.wikipedia.org/w/index.php?title=File:Flag_of_Sri_Lanka.svg *License*: unknown *Contributors*: Zscout370

File:Flag of Vietnam.svg *Source*: http://en.wikipedia.org/w/index.php?title=File:Flag_of_Vietnam.svg *License*: unknown *Contributors*: user:Lưu Ly

File:Flag of Indonesia.svg *Source*: http://en.wikipedia.org/w/index.php?title=File:Flag_of_Indonesia.svg *License*: unknown *Contributors*: User:Gabbe, User:SKopp

File:Flag of Pakistan.svg *Source*: http://en.wikipedia.org/w/index.php?title=File:Flag_of_Pakistan.svg *License*: unknown *Contributors*: Zscout370

File:Flag of Algeria.svg *Source*: http://en.wikipedia.org/w/index.php?title=File:Flag_of_Algeria.svg *License*: unknown *Contributors*: User:SKopp

File:Flag of Iran.svg *Source*: http://en.wikipedia.org/w/index.php?title=File:Flag_of_Iran.svg *License*: unknown *Contributors*: Various

File:Flag of Kazakhstan.svg *Source*: http://en.wikipedia.org/w/index.php?title=File:Flag_of_Kazakhstan.svg *License*: unknown *Contributors*: -xfi-

File:Flag of Azerbaijan.svg *Source*: http://en.wikipedia.org/w/index.php?title=File:Flag_of_Azerbaijan.svg *License*: unknown *Contributors*: User:SKopp

Image:Economist Intelligence Unit.png *Source*: http://en.wikipedia.org/w/index.php?title=File:Economist_Intelligence_Unit.png *License*: unknown *Contributors*: User:Coat of Arms, User:Cydebot

Image:2005ICT.PNG *Source*: http://en.wikipedia.org/w/index.php?title=File:2005ICT.PNG *License*: unknown *Contributors*: Mdd, Shushruth, 1 anonymous edits

Image:World Bank Logo.svg *Source*: http://en.wikipedia.org/w/index.php?title=File:World_Bank_Logo.svg *License*: unknown *Contributors*: Godvarryu, Ichwan Palongengi

File:WhiteandKeynes.jpg *Source*: http://en.wikipedia.org/w/index.php?title=File:WhiteandKeynes.jpg *License*: unknown *Contributors*: International Monetary Fund

File:World Bank building at Washington.jpg *Source*: http://en.wikipedia.org/w/index.php?title=File:World_Bank_building_at_Washington.jpg *License*: unknown *Contributors*: Avicennasis, Dodo, DrJunge, FAEP, Wouterhagens, 3 anonymous edits

File:Justin Yifu Lin 1-2.jpg *Source*: http://en.wikipedia.org/w/index.php?title=File:Justin_Yifu_Lin_1-2.jpg *License*: unknown *Contributors*: Zolo,

File:Energy Star logo.svg *Source*: http://en.wikipedia.org/w/index.php?title=File:Energy_Star_logo.svg *License*: unknown *Contributors*: Fred J, Rocket000, Tetris L, 1 anonymous edits

Printed by Books on Demand GmbH, Norderstedt / Germany